Political Communication in Canada..........

**COMMUNICATION
STRATEGY
AND POLITICS**

Communication, Strategy, and Politics

THIERRY GIASSON AND ALEX MARLAND, SERIES EDITORS

Communication, Strategy, and Politics is a ground-breaking series from UBC Press that examines elite decision making and political communication in today's hyper-mediated and highly competitive environment. Publications in this series look at the intricate relations between marketing strategy, the media, and political actors and explain how this affects Canadian democracy. They also investigate such interconnected themes as strategic communication, mediatization, opinion research, electioneering, political management, public policy, and e-politics in a Canadian context and in comparison to other countries. Designed as a coherent and consolidated space for diffusion of research about Canadian political communication, the series promotes an interdisciplinary, multi-method, and theoretically pluralistic approach.

Other books in the series are:

Political Marketing in Canada, edited by Alex Marland, Thierry Giasson, and Jennifer Lees-Marshment

Political Communication in Canada..........

Meet the Press and Tweet the Rest

...... Edited by
Alex Marland, Thierry Giasson,
and Tamara A. Small

UBCPress · Vancouver · Toronto

22 21 20 19 18 17 16 15 14   5 4 3 2 1

Printed in Canada on FSC-certified ancient-forest-free paper
(100% post-consumer recycled) that is processed chlorine- and acid-free.

Library and Archives Canada Cataloguing in Publication

     Political communication in Canada : meet the press and tweet the rest /
edited by Alex Marland, Thierry Giasson, and Tamara A. Small.

(Communication, strategy, and politics, ISSN 2368-1047)
Includes bibliographical references and index.
Issued in print and electronic formats.
ISBN 978-0-7748-2776-8 (bound). – ISBN 978-0-7748-2777-5 (pbk.)
ISBN 978-0-7748-2778-2 (pdf). – ISBN 978-0-7748-2779-9 (epub)

     1. Communication in politics – Canada. 2. Mass media – Political aspects – Canada.
3. Social media – Political aspects – Canada. 4. Advertising, Political – Canada. 5. Journalism
– Political aspects – Canada. 6. Political campaigns – Canada. I. Marland, Alexander J.,
editor II. Giasson, Thierry, editor III. Small, Tamara A. (Tamara Athene), editor IV. Series:
Communication, strategy, and politics

JA85.2.C3P635 2014   324.7'30971   C2014-904879-3
                                                      C2014-904880-7

Canadä

UBC Press gratefully acknowledges the financial support for our publishing program of the Government of Canada (through the Canada Book Fund), the Canada Council for the Arts, and the British Columbia Arts Council.

This book has been published with the help of a grant from the Canadian Federation for the Humanities and Social Sciences, through the Awards to Scholarly Publications Program, using funds provided by the Social Sciences and Humanities Research Council of Canada.

UBC Press
The University of British Columbia
2029 West Mall
Vancouver, BC V6T 1Z2
www.ubcpress.ca

Contents

.... # Figures and Tables

Preface

This book is a follow-up, of sorts, to *Political Marketing in Canada* (Marland, Giasson, and Lees-Marshment 2012), which argued that the main reason that Canadian political elites use market intelligence such as opinion polling and focus group data is to inform their communication decisions. In *Political Communication in Canada*, we explore ways that changes in communication technology and media behaviour are affecting Canadian politics. This includes the communication between political parties, politicians, public servants, interest groups, the media, and Canadian citizens in the digital age.

In the preface to *Political Marketing in Canada,* Conservative Party marketer Patrick Muttart is noted as emphasizing that political parties manage earned media, paid media, direct voter contact, local matters, and social media simultaneously during an election campaign. This requires the use of centralizing tools. The Conservatives use their Constituency Information Management System (CIMS) computer software to organize information about electors, while within the Government of Canada they introduced the Message Event Proposal (MEP) to coordinate thematic messaging. As the news cycle speeds up, and as the line between an election campaign and inter-election period blurs, this media management is now constant. Brad Lavigne, who was the New Democratic Party's director of strategic communications during Jack Layton's tenure and the NDP campaign director in 2011, advises that communication management has taken on greater importance in politics for the following reasons:

1   An increasingly persuadable electorate: as voters become less entrenched in traditional voter behaviour (based on family, geography, religious, or class), Lavigne says, they are increasingly open to switching their party preference and therefore susceptible to political communication;

2   The constant campaign: the accelerated number of elections in short succession (four federal elections between 2004 and 2011), along with the introduction of the "permanent campaign" has forced political parties to be constantly engaged, allowing for expertise and new tools to develop; and,

3   Platform proliferation: The explosion of media platforms with which political parties can get their message out, including twenty-four-hour Canadian cable news channels, all-news radio, online news websites, and social media, has created a seemingly limitless marketplace for political parties to communicate with an ever-increasing persuadable audience (Lavigne 2013).

The strategic approach of such experienced practitioners is rooted in truisms of political science and of political communication. As partisanship declines the number of floating voters is increasing; the proximity of an election moves parties into campaign mode; and persuasive messaging is constantly communicated through a multitude of media. What this means for Canadian democracy is that campaigners' intensifying attitude to political communication is persisting after the election and into governance. Data are being used to inform the targeting of segments of floating voters. Narrow messages reach select citizens via specialized media. The mediation function of political journalists is bypassed, and people who support other parties are left out. A subtext is that despite the engagement opportunities of Web 2.0, the priority remains on getting the party line out. Hence the subtitle of this book, *Meet the Press and Tweet the Rest* – though even personal interactions between politicians and the press are in flux and are increasingly mediated by digital media. To date, Canada's major parties have decided that the benefits of interactive political dialogue are not worth the risk of losing control of the message.

We believe that this changing communications environment merits study. This book was organized by two of the editors of *Political Marketing in Canada* and one of the chapter authors. It brings together a variety of scholars, including some who had previously researched political marketing and others whose area of expertise is political communication. A project such as this represents a collective effort. It was a pleasure to work with authors who were so responsive to feedback, who delivered material within stated timelines, and who were themselves good communicators. The editors would like to thank Memorial University research assistants Lori-Ann Campbell,

Kayla Carroll, Michael Penney, and Matthew Yong, who provided capable and timely copyediting of the draft manuscript. We appreciate the professionalism exhibited by staff at UBC Press, in particular Megan Brand, David Drummond, Valerie Nair, and Judy Phillips. A special thanks to senior editor Emily Andrew who, in addition to providing capable and timely support, has encouraged the development of the Communication, Strategy, and Politics series. We would also like to thank the three anonymous reviewers who, through UBC Press, provided helpful feedback on an earlier draft of the manuscript.

The editors wish to acknowledge financial assistance that has made this book possible. *Political Communication in Canada* has been published with the help of a grant from the Canadian Federation for the Humanities and Social Sciences, through the Award to Scholarly Publications Program, using funds provided by the Social Sciences and Humanities Research Council of Canada. A financial award was granted from Memorial University of Newfoundland's Publications Subvention Program, through the Office of Research Services. Finally, the Centre for the Study of Democratic Citizenship provided funding that supported a workshop on political communication that was coordinated by the editors at the 2012 Canadian Political Science Association annual conference, held at the University of Alberta. A number of contributors presented their work there, which featured a discussion panel of political communication scholar David Taras of Mount Royal University and Dimitri Soudas, a former director of communications in Stephen Harper's Prime Minister's Office.

Political Marketing in Canada was well received as a volume that has raised awareness of the growing practice of marketing in Canadian politics. It is our hope that *Political Communication in Canada: Meet the Press and Tweet the Rest* contributes to scholars', students', and practitioners' interest and understanding of the use of communication in the Canadian political realm.

Alex Marland
Lead Editor

PART 1

Communication by Canadian Political Institutions . . . .

1

The Triangulation of Canadian Political Communication

Tamara A. Small, Thierry Giasson, and Alex Marland

"Just a man and his pen." This is how a *Maclean's* article described Canada's prime minister in 2011 (Wherry 2011). Whereas US president Barack Obama fought to keep his BlackBerry when taking office, Prime Minister Stephen Harper has apparently never owned a smartphone. The implication is that in an era of email, texting, and video calling, the most powerful elected official in Canada continues to opt for a pen and paper. This anecdote seems at odds with the personal experience of many Canadians for whom life without access to the Internet, let alone a smartphone, has become unthinkable. Digital technology has infiltrated daily life in so many ways. It has changed personal communication, business and commerce, education, and, as this book will show, Canadian politics.

Although Canadian political communication tends to be a technological laggard compared with the United States, it nevertheless evolves over time. Without technology, politicians are confined to personal communication, such as delivering local speeches and meeting electors on their doorsteps. With technology, politicians can communicate with larger audiences, using rapid transportation and mass communication. As discussed in this book's predecessor, *Political Marketing in Canada* (Marland, Giasson, and Lees-Marshment 2012) and elsewhere (Flanagan 2007; also Chapter 14 in this book), computer-assisted research enables parties to use data to segment the electorate, to target messages using narrow appeals, and to make market-based decisions about selecting specialized media. In theory, political marketers avail of new technology to respond to public opinion in a manner that enhances democracy because the general public's concerns are prioritized over those of political elites. But in practice, political elites tend to use opinion data and marketing to inform decisions that prioritize their own interests. Political communication is more straightforward. It is concerned with mediated and unmediated interactions between political elites and citizens. The digital era has been a transformative period for communication because the Internet has empowered the masses to join a cadre of political and media elites as content creators and as information disseminators. However, like the paper and pen, even in the new media environment, the tried-and-true methods of communicating politically – whether through press conferences, direct mail, television advertising, or even door knocking – still matter. Moreover, regardless of the medium, the prevalence of partisanship can supersede idealistic notions of democratic discourse, much as it did in past centuries.

It is here that *Political Communication in Canada* is positioned: at the intersection of politics operating in a traditional media environment that is adapting to significant technological change. This book explores the range of political communication activities used by Canadian political institutions, the mass media, and citizens. Some chapters focus on the current state of traditional communication activities, including advertising and media management; others explore newer digital technologies, such as blogging and Twitter, as well as strategic considerations such as political branding and personality politics. By looking at political communication in the entirety of its process and from the perspective of a variety of actors, *Political Communication in Canada* fills a current knowledge gap in Canadian scholarly literature. It begins here by providing a conceptual foundation that includes defining political communication, exploring how political communication changes, and summarizing the key research themes.

What Is Political Communication?
The essence of politics is talk or interaction (Denton Jr. 2009, xiii). Regardless of the mode, whether a speech at a political rally, a televised debate between leaders, door-to-door canvassing, a media interview, or a Twitter post by an MP, communication provides a link between those in power and citizens, and among citizens. In all of these cases, political actors are talking about politics with the intention to inform, persuade, promote, and even influence their audience's behaviour.

Broadly defined, political communication can be thought of as the "role of communication in the political process" (Chaffee 1975, 15). It can take place in a variety of forms (formal or informal), in a variety of venues (public and private), and through a variety of mediums (mediated or unmediated content). It includes the production and generation of messages by political actors, the transmission of political messages through direct and indirect channels, and the reception of political messages. There is a tendency to explore political communication as a process that flows out from political institutions, downward from institutions to citizens (Lilleker 2006). We, however, see it as a triangular process that includes political institutions and actors, the news media, and, importantly, citizens. Every act of political communication produced by parties, interest groups, or the media is geared toward citizens, to inform them, to influence them. It is the interactions between these three groups that matter in political communication. Within politics, communication flows move in many directions: downward from governing authorities to citizens; horizontally between political actors, including news media; and upward from citizens and groups to the political institutions (Norris 2001).

As discussed, the world of political communication is changing as political actors adapt to technological innovations. Scholars have divided the history of political communication into various stages or eras (Blumler and Kavanagh 1999; Farrell and Webb 2000; Norris 2000; Gibson and Römmele 2001). In the first stage, political communication was organized around direct, face-to-face communications such as canvassing, meetings, and rallies. For instance, Prime Minister John A. Macdonald instituted the first "campaign picnic" as a method of political communication in the election of 1878, with the goal of presenting himself as an approachable man rather than a distant politician (Nolan 1981). Nolan notes that, in addition to the campaign picnic, political communication in early campaigns revolved around newspapers and public meetings and then later the campaign train and radio. The ferocity of the partisan press was such that it was not uncommon for Macdonald to strategize about buying newspapers in communities where his party and the government needed to shore up support (Levine 1993).

The advent of broadcast media, notably television, brought forth a new era of political communication. Whereas partisan and commercial interests prevailed in the press, the creation of a public broadcaster, Canadian Broadcasting Corporation (CBC)/Société Radio-Canada (SRC), in the 1930s sought to offer mediated balance and a Canadian perspective. Not only did radio and television enlarge the audience, but the emergence of the latter in the 1950s brought new dimensions to politics, including a predisposition toward the visual, the dramatic, and mass appeal (Taras 1990). How well political actors can perform on and manipulate television has become just as important as what is being said. At the same time, the press has evolved to become less partisan, which has led to political actors competing for media attention across various mediums. As a result of these factors, media management, polling, and professionalization became the innovative tools of the trade for political actors during the era.

There is general agreement that a new political communication age has recently developed. This is characterized by factors such as a multiplicity of channels and technologies, and a 24/7 information environment. Television remains important, but a few main channels no longer dominate. Cable and satellite technology have extended the options available to viewers, which results in audience fragmentation. Digital technologies, including the Internet, mobile technologies and related applications, have furthered this fragmentation. Farrell and Webb (2000) suggest that these new direct modes of communication are increasingly given more weight by political actors. This diversification of both information and audience has implications for political communication. The once powerful news media continues to be a trusted source of information for most Canadians (CMRC 2011c) but is facing increased competition from specialty channels, alternative news outlets, blogs, and social media (e.g., Kozolanka, Mazepa, and Skinner 2012). For political actors, it is easier to communicate and yet more difficult to get political messages out to a diffuse audience within a diversified media environment. Blumler and Kavanagh (1999, 213) describe this new media system as a "hydra-headed beast" with "many mouths ... which are continually clamoring to be fed." As such, professionalization, political marketing, and targeted communications are necessary tools for getting political messages to citizens more effectively. Moreover, citizens have access to new media in a way they did not in the previous era, and are able to communicate with elites and share information with each other in ways not previously possible.

Although it is clear that political communication has changed, each new era does not completely displace the previous one. Television advertising has not replaced print or radio ads, and the Internet has replaced none of these. Rather, the growing number of new media platforms and specialty outlets increases the variety of contact points with target audiences. The need to integrate various communications is perhaps shown best in the 2012 US presidential election, where despite being crowned the "Internet president," Barack Obama had a considerable ground campaign in the battleground states where door-to-door and telephone canvassing were critically important (Gabriel 2012). No country has completely transitioned into a post-television stage, but a new communication context is emerging in many advanced industrial countries, including in Canada. And if politics takes place through communications, it is important that we understand how this new context shapes and changes Canadian politics and democracy.

The Study of Political Communication in Canada

As a field of study, political communication is interdisciplinary, bringing together a diversity of methodological and theoretical approaches. It is as old as politics itself, though academic interest is more recent (Lilleker 2006), especially in Canada. Indeed, political communication studies have evolved significantly in the last two decades, and interest in teaching and researching this field is therefore both constant and growing. In recent years, special panels and workshops have been organized at Canadian communication and political science associations' annual meetings dealing with topics such as news media, Web 2.0, and political communication. Across Canada, research teams are dedicating significant attention to the study of aspects of political communication, such as the Canadian Election Study, the McGill Institute for the Study of Canada, the Montreal-based Centre for the Study of Democratic Citizenship, the Groupe de recherche en communication politique (Université Laval), the Infoscape Research Lab (Ryerson University), the Canadian Media Research Consortium (University of British Columbia), and the Canada Research Chair in Electoral Studies (Université de Montréal).

The bulk of Canadian political communication research finds its roots in liberal democratic theory. As a reflection of the scholarly and public interest in political communication, Canadian scholars have been publishing numerous edited collections (e.g., Romanow et al. 1999; A.-M. Gingras 2003; Taras, Pannekoek, and Bakardjieva 2007; Sampert and Trimble 2010; Kozolanka, Mazepa, and Skinner 2012) and monographs (e.g., Nesbitt-Larking 2001; Taras 2001; Soroka 2002; Miljan and Cooper 2003; Gidengil et al. 2004; A.-M. Gingras 2009; Fleras 2011; Taras and Waddell 2012a) dealing with the association of politics, media coverage, and citizenship in Canada. These scholars tend to look at the instrumentalization of communication for electoral purposes, the functions of news media in Canadian democracy, the rational use of political information by voters, and the impact of political communication messages on attitude formation or levels of political participation.

This book continues in the liberal democratic tradition, though many scholars, including Canadians, look at political communication objects through critical lenses. Within the political economy approach, research explores how the structures of ownership and control of media organization, advertising, and legal regulations shape the production of news (see, for example, Taras 2001; Skinner, Compton, and Gasher 2005; A.-M. Gingras 2009). Elite theorists, for instance, argue that media owners collaborate with the corporate and governing elites in news production in order to maintain class dominance. As we will see in the discussion of the news media, this research explores blind spots in the media toward gender, minorities, inequality, poverty, and labour (Hackett 2005). Other scholars, interested in reflecting on the societal, political, and democratic implications of hegemonic discourses, myths, or representations, look at the transformations in communication activities by parties, citizens, and organizations through culturalist theories. Inspired by cultural studies and French political sociology, they study how political communication practices and technologies convey and reinforce social norms and order as ideological tools from dominant political structures (see, for example, Howard 2006; Proulx and Kwok Choon 2011; Proulx 2012). These important reflections highlight how communication activities and technologies can be used by actors to limit democratic practices, such as voting or accessing information, and how they attenuate the private sphere but also the way civil society can harness political communication for resistance purposes to discrimination or alienation.

With a few notable exceptions (Gidengil et al. 2004; Taras, Pannekoek, and Bakardjieva 2007; Taras and Waddell 2012a), published works in Canada have mostly looked at the intersection of partisan politics and traditional media coverage, often in an electoral context. Yet political communication as a field of study and practice is broader than this. For instance, there seems to be somewhat more limited empirical research dedicated to the reception aspects in Canada, and less is known about how Internet technologies are used. By providing readers with an analysis of the three constituents of political communication, and by focusing on the ways that recent technological and societal transformations have modified how and by whom it is produced and disseminated, this book represents a body of knowledge that builds on international and Canadian scholarly literature of this evolving era of communicating the political.

The Triangulation of Political Communication: Conceptual Nuances and Distinctions

Before we can analyze political communication in Canada, we need to establish conceptual nuances by exploring the key activities of the three constituents of political institutions, the news media, and Canadian citizens. Although an exhaustive literature review is not possible here, this section highlights some key research themes in these three areas, with special attention paid to contributions by Canadian scholarship.

COMMUNICATION BY CANADIAN POLITICAL INSTITUTIONS

Political institutions, including political parties and leaders, the public service, parliamentarians, the judiciary, interest groups, and non-governmental organizations, are important political communicators. Such actors communicate their activities in order to gain legitimacy and compliance of citizens (Lilleker 2006). We suggest that communication by these political actors in Canada has several characteristics.

First and foremost, the communication of Canadian political institutions is marked by the intensity of partisanship. The pressure to conform, the salience of rewards, and the fear of sanctions act as a communications glue. This cohesion (e.g., Chris Kam 2001) ensures that public remarks are consistent with a political organization's official position, which is commonly referred to as being "on message." It is rare for individuals to publicly state a position that is at odds with that of a designated spokesperson; therefore, many partisans avoid interacting with the media until they can repeat what the spokesperson communicated.

This message discipline seems to have reached new heights with the Conservative Party under the leadership of Stephen Harper. As some of the chapters in this book describe, Harper's team has sanctioned reduced interactions with the Canadian Parliamentary Press Gallery in favour of communicating with small media outlets; has restricted MPs' and public servants' media availability; has replaced Liberal red on government websites with Tory blue; and has micromanaged departmental communications through the introduction of Message Event Proposal forms demanded by central agencies (see also Kozolanka 2009, 2012). Such communication centralization ensures message consistency and helps the governing party influence the public agenda as the media environment intensifies and becomes more fractured. However, it raises questions about the nature and extent of democratic discourse in this country.

Second, there are temporal dimensions that affect communication in Canadian politics. Even in a 24/7 media world, the news cycle continues to be organized around the evening television news, though there is increasing pressure to report news as it happens. The amount of media attention paid to politics is related to the parliamentary cycle; for instance, during the budget period there is intense national public interest as compared with the summer months, when the legislature is not in session. The most concentrated period of political communication occurs during an election campaign, when choosing a party and leader to head the national (or provincial) government is at the top of the public agenda. Predictable variations in communication occur during periods of public celebration, mourning, and days of rest, but what tend to be unpredictable are the timing and nature of a crisis. Canada's political communicators attempt to structure their decisions around these ebbs and flows.

Third, regional dimensions result in variations in communications strategies and tactics. Mass communication is obviously an essential political tool in a country with a geography that is as vast as Canada's. But what is communicated from one part of the country may not resonate uniformly within that jurisdiction, let alone across the entire nation. Canadians have different value systems, which often depend on whether they reside in urban, suburban, or rural areas. Most significant are the variances between provinces and regions, which each have their own political culture, institutions, and media systems (e.g., Ornstein, Stevenson, and Williams 1980). In terms of political campaigns, Carty, Cross, and Young (2000) suggest that regionally targeted advertising is a necessity because of the need to maximize finite campaign resources. Parties need to spend money efficiently, which means targeting their communication in clusters of electoral districts where the outcome is in question, and paying less attention to areas where the election result is predictable. Political parties also have a history of communicating different or even conflicting messages in different regions, especially in Quebec (e.g., Neatby 1973), though the rapidity of information flows is curtailing this practice.

Fourth, political communication in Canada operates in a bilingual sphere that is increasingly multilingual. Section 16 of the Charter of Rights and Freedoms directs that the executive, legislative, and judicial branches of the federal government, including the public service, are required to communicate in both official languages (Canada 2012a). Furthermore, campaign advertising, party documents, and many online materials tend to be produced in English and French. However, it is the increasing number of allophones – Canadians whose native language is neither English nor French – that is spurring change in how political institutions communicate. According to census data, the proportion of Canadians whose mother tongue is French is comparable to the growing number of allophones, most of whom live in metropolitan areas (Canada 2011). Among those speaking a non-official language, Chinese, including Cantonese and Mandarin, is today by far the most common language, followed by Italian, German, Punjabi, Arabic, Tagalog (Filipino), Portuguese, and Polish, among others, such as mother tongues spoken by Aboriginal peoples. Political communicators are increasingly reaching out to these citizens in their own language and via specialty media outlets. This includes the federal government, for section 27 of the Charter provides for "the preservation and enhancement of the multicultural heritage of Canadians," as well as political parties. For instance, the federal New Democratic Party (NDP) has produced election campaign materials in Chinese languages, Korean, and Punjabi (Whitehorn 2006). The Government of Canada spends hundreds of thousands of dollars annually to monitor so-called ethnic media (CBC News 2012b), and 4.2 percent of its 2010-11 advertising budget was on community media that serve ethnic minority communities (calculated from Canada 2012b).

Fifth, rules of the game affect political communication in Canada, as they do with the practice of political marketing (Dufresne and Marland 2012). This is especially true during elections, which in Canada are heavily regulated, if poorly enforced. The official campaign period is subject to the provisions of the Canada Elections Act, which limits fundraising and spending, subsidizes both through rebates, and provides free broadcast time in the name of levelling the playing field for election communication. Outside of elections or leadership contests, however, few rules apply, and political actors are constrained foremost by their financial situation and the barometers of good taste.

Sixth, advertising has a prominent place in Canadian political communication, not only because of its persuasive ability but also because of its sponsor's capacity to control the message and reach target audiences. The Canadian Code of Advertising Standards promotes principles such as truthfulness and message accuracy. Its non-binding provisions apply to governments and government departments, as well as to Crown corporations; however, the code exempts political and election advertising lest it impinge on "the free expression of public opinion or ideas" (Advertising Standards

Canada 2012). Whereas political institutions are normally subject to journalistic interpretation, advertising allows them to frame themselves, their issues, and their opponents in their own ways directly to citizens. There is, however, a significant need for resources to have this level of control. Advertising takes time to create and requires specialist expertise, there are a multitude of media platforms to choose from, and above all, it is expensive. To illustrate, in 2012, the price of a single full-page colour ad in *Maclean's* was $38,940 (*Maclean's* 2012), and from 2006 to 2011, the Government of Canada spent an average of $94 million annually on advertising campaigns, including those about the H1N1 flu pandemic, the health and safety of children, elder abuse, credit card regulations, economic stimulus spending, and tax credits (Canada 2012b). Although Internet spending had grown to 15 percent of all government advertising expenditures by 2010-11, 72 percent was on traditional broadcast and print media, with TV spots commanding nearly half of all monies spent (ibid.).

Government and political party advertising invite scrutiny. With government advertising, the concern is that cabinet is responsible for government strategy and expenditures, including steering the direction of communication. Critics fear that a governing party will use public resources to create partisan propaganda in an effort to set the public agenda and build support for their political priorities (e.g., Ellul 1965). As well, advertising agencies that help a political party win an election campaign may have a quid pro quo expectation that they will receive lucrative government advertising contracts in return. In Canada, there is a long-standing history of what Rose (2000, 89) refers to as the "symbiotic relationship" between wealthy parties and advertising professionals. One of the biggest political scandals in Canadian history involved ad agencies affiliated with the Liberal Party of Canada obtaining public funds by submitting inflated or fake invoices that were paid by the Liberal government (Kozolanka 2006). Rose (2000) has also found that government advertising campaigns can be controversial because they are used to support policy agendas. Recently the Conservative administration has been accused of financing government advertising that conveys its party's values and priorities, such as Canadian Forces recruitment, the economic action plan, and the War of 1812 campaign (e.g., Kozolanka 2012). Fortunately, transparency and rules are improving. Advertising contracts are routinely put out to public tender; the federal government issues annual reports on advertising expenditures; and businesses are no longer allowed to donate to federal parties. Still, there is no federal advertising review board to ensure the suitability of government advertising as there is in Ontario (see Office of the Auditor General of Ontario 2012a).

With respect to advertising paid for by political parties, the Canadian literature often focuses on the use of attack or negative ads. That these are often called "American-style" ads implies that Canadian political culture is different and that it imports political communication tactics, including negativity (Rose 2004). Rather than communicate information about a candidate's strengths and merits, negative ads emphasize what is wrong with opponents personally and with their policies, often in a sinister tone (e.g., Kaid 2000). Negative ads have figured significantly in Canadian election campaigns, with the most notorious case being the 1993 "face" attack ad that brought public attention to Liberal leader Jean Chrétien's partial facial paralysis (Romanow et al. 1999; O'Shaughnessy and Henneberg 2002). The intensity of partisan advertising in the minority government era from 2004 to 2011 demonstrates that although party operatives learned from the 1993 incident not to attack personal appearances, they nevertheless continue to communicate negative messages. For instance, the Conservative Party was relentless in its inter-election advertising that critiqued Liberal leaders, and over half (58 percent) of television spots in the 2008 general election were negative (Rose 2012). This too invites debate on the implications for the quality of democratic discourse in Canada. On one hand, negative ads are said to be responsible for suppressing elector turnout and, as in 1993, they can anger the electorate (Cunningham 1999). Yet others maintain that democracy "requires negativity" by promoting key democratic values such as opposition and accountability (Geer 2006, 6). Regardless, the ongoing and growing expense of political communication has the parties looking for competitive advantages in fundraising, in audience targeting, and in achieving earned media coverage – without losing control of the message.

Finally, Canadian political institutions have adapted to the growing prevalence of online media, but their preference for one-way communication with citizens has persisted. The federal e-government strategy has been successful in terms of service delivery (Longford 2002), and parliamentarians see information communication technologies as a benefit to their roles as representatives (Kernaghan 2007). However, scholars question whether this has contributed to a more participatory ethos (see Roy 2006). Parliamentarians are overloaded by email, and they tend to avoid interactive technologies such as online chats or blogs (Kernaghan 2007; Small 2008a). During federal election campaigns, there has been minimal online interaction between citizens and political parties (Small 2004, 2008b, 2010b; Francoli, Greenberg, and Waddell 2012). In political circles, the Internet tends to be used for traditional campaign purposes, including for communication, fundraising, and organizing, rather than for two-way dialogue. In all cases, Canadian political institutions have expended considerable money and effort on their online presences, and use them to provide information and services to citizens. However, much of this continues to be one-way communication, with little evidence of a paradigmatic shift in their communication activities because of the Internet.

Canadian Political News Media

Nestled between political institutions and Canadian citizens is the news media. The democratic importance of the media can be seen in the constitutional entrenchment of press freedoms in section 2 of the Charter. In terms of political communication, the media perform two key public service roles: educators and watchdogs (Fletcher 1981). Much of what citizens know about politics comes from the mass media. Most Canadians will never speak directly to a political leader, attend a political event, or read the text of a statute. It is through the news media that citizens engage with politics and come to form opinions about it. The media is also the watchdog of the political system. Through organizations such as the press gallery and investigative journalism, the media ensure the legitimacy of the Canadian political system by holding political actors to account for their actions and behaviours. Reporters and journalists, therefore, provide a necessary link between citizens and political elites. These are among the reasons that the news media is considered an "essential pillar of democracy" (Siegel 1993, 18).

Hackett (2001) identifies five enduring characteristics of the Canadian media. First, as noted, through section 2 of the Charter, the Canadian media is independent from the state. Second, unlike Europe, the Canadian media is non-partisan, though this does not necessarily mean unbiased. As mentioned, this was not always the case, for newspapers were once vehicles of partisan propaganda, with most major centres having two papers, each of which represented one of the two main parties (Taras 1990; Levine 1993). Social and technological changes led to a commercial press, where objectivity, rather than partisanship, became the mass appeal of the news organization. Subsequently, the partisan press died out by the 1960s (Taras 1990). However, in the new communication environment, the consumption of information communicated in a printed format is on the decline and newspaper circulation in Canada is sliding (Androich 2011). Third, the Canadian media reflects the bilingual diversity of Canada. Fletcher (1998) points out that French- and English-language media operate in separate worlds. French- and English-language public affairs reporting share relatively little in common. Moreover, as noted, the news media is increasingly reflecting Canada's multilingual diversity. Fourth, much of the Canadian media is privately owned, and because of legal restrictions, these owners are predominately Canadian. As private businesses, the media seeks to make a profit, and related to this is the concentration of media ownership. According to Nesbitt-Larking (2001), Canada has one of the highest concentrations of media ownership among capitalist countries. As of 2012, eight corporations owned the majority of Canada's media (Theckedath and Thomas 2012), which raises concerns about media concentration and points to the important role of public broadcasting and the Internet.[1] The fifth characteristic of the Canadian media is the presence of the CBC/SRC, which was created to protect Canadian culture from American influence. Its role as a broadcaster is morphing as technology and media economics change. One of the CBC's current priorities is "to make significant investments in its online and digital platforms to bring programming and services to more Canadians in new ways" (CBC Radio-Canada 2013). Unlike its Canadian media competitors, which are increasingly restricting access to their online content through metered paywalls and subscriber-only services, the CBC has a growing array of unlocked online content. Yet the CBC is itself under financial pressure, and its ability to communicate Canadian political information on radio and TV is constrained by serious budgetary issues (Friends of Canadian Broadcasting 2011). Taken together, these five characteristics shape the operation of the mass media in Canada, which in turn shapes political communication in this country.

How Canadians get their news is evolving in the new media environment. A 2011 survey found that over 80 percent of Canadians felt that mainstream media were reliable sources of information; by comparison, about 40 percent trusted government information and 25 percent felt that social media was trustworthy (CMRC 2011c). Source credibility is a significant reason the majority of Canadians preferred to get their news from mainstream media platforms (38 percent TV, 23 percent newspapers, 8 percent radio), compared with 31 percent who preferred the Internet (CMRC 2011b). However, the survey found significant perception differences among Canadians aged eighteen to thirty-four, who placed more credibility on social media than did older Canadians. Moreover, those who use social network sites often do so to gather news, and thus it is estimated that over 10 million Canadians enjoy a "personalized news stream" (CMRC 2011d). Further, more than half (52 percent) of Canadians said that the Internet was a source of the most interesting news and information, compared with 27 percent citing television, 15 percent newspapers, and 6 percent stating radio (CMRC 2011b). This presents a significant challenge for organizations seeking to present news gathered by professionals given that only 4 percent of Canadians were willing to pay for news online (CMRC 2011a). Television may remain the main source of news for Canadians, but information that is available for free online has become a serious competitor.

Another aspect of this freebie culture is the popularity of free newspapers such as *Metro* or *24 Hours* (Sauvageau 2012, 32). Found in large cities worldwide, free newspapers are primarily targeted at commuters. There were twenty-nine free dailies printed in Canada in 2012, including in Toronto, Vancouver, Montreal, and Calgary (Baluja 2012). The Toronto versions of these papers reach about a quarter of a million people each day. Sauvageau (2012) suggests that free newspapers are very popular among younger people, which seems at odds with predictions of the demise of the newspaper industry. Nevertheless, free dailies, along with free online media, contribute to substantial financial problems for traditional media organizations in Canada and elsewhere.

Given the importance of the fourth estate in democracy, Canadian scholars have given considerable attention to the news media and their political role. Media frames are an important research theme. According to Gitlin (1980, 7), media frames are "persistent patterns of cognition, interpretation, and presentation, of selection, emphasis, and exclusion, by which symbol-handlers routinely organize discourse, whether verbal or visual." These are interpretative cues used by journalists that give meaning to issues. In the game frame, also known as the horserace or strategy frame, campaigns are conceived as strategic races between political actors. This is common in Canada, as the media focus on who is ahead and who is behind in the polls, rather than on the substance of the issues (Taras 2001, 148; Giasson 2012; also Ansolabehere and Iyengar 1994).

Related to the issue of horserace framing is polling. Central to the "who's in front" or "who's falling behind" narratives, polls have become a permanent fixture of media coverage. Indeed, Turcotte (2012) estimates that a new poll was released every three days between the end of the 2008 federal election and the signing of the writ in 2011, and sixty-seven polls were released during the campaign itself. The media commissions many of these polls and are often formally attached with polling companies. For instance, during the 2011 election, Ipsos conducted polls for Postmedia News and Global National, while CBC worked with Environics. Since polls often become newsworthy, pollsters work with and for the media to enhance their own corporate reputation (Adams 2010).

Often subject to criticism, election polls have recently come under fire for their inability to predict election outcomes and thus their purported lack of accuracy. In recent provincial elections in Alberta and British Columbia, pollsters predicted landslide victories for the opposition, when in fact the governing parties were re-elected. In British Columbia in 2013, many polls released immediately prior to Election Day indicated that the NDP enjoyed a lead of eight or nine points over the Liberals. On election night, however, the Liberals earned a five-point victory over the NDP and won five more seats than they had in the previous election. As with the 2012 Alberta election, pollsters were lambasted by the media and in social media (*Huffington Post BC* 2013). The polling industry defended their results but nevertheless acknowledged that the industry is facing challenges because of evolving communications technology. Some methodological concerns for pollsters include the emergence of Internet polling and political weighting of declining turnout (Pickup et al. 2011). Some believe that the reputation of the polling industry has been damaged by these recent missteps (Grenier 2013; A. Reid 2013). Reaching electors by telephone is also a challenge, given that most cell-phone numbers are unlisted and a growing proportion of the population is opting not to have a landline.

Another trend in political coverage is related to tone. Scholars have found that the tone of political media reports in Canada has become increasingly negative over the past four decades (Nevitte et al. 2000, 24). For instance, Trimble and Sampert (2004, 55) found that there were more positive than negative remarks about political actors in 1962 and 1974, but by the 1979 federal election, the media's tone began to change, with negativity in the media prevailing over positive coverage in the 1980, 1984, 1993, and 2000 elections. Such findings support a position that media coverage is no longer deferential to elites and that journalists are acting as government watchdogs. Yet they can also be used to support arguments that media reports are slanted and excessively anti-establishment.

The potential for media slants leads to the study of norms and assumptions that are found in editorial decisions about news selection, emphasis, and framing. In an all-encompassing analysis, *The Media Gaze* (Fleras 2011) found differences in the media representation of gays and lesbians, women, youth, Aboriginal people, radicalized minorities, and religion. A wider body of gender-mediation research has examined the limited news coverage of female politicians, the lack of neutral coverage, and stereotyping (F.-P. Gingras 1995; Robinson and Saint-Jean 1995; Trimble and Everitt 2010; Lalancette and Lemarier-Saulnier 2013). Since politics has been and is still dominated by men, it is often framed in a masculine narrative (e.g., sporting and war metaphors), though overt gender stereotyping is on the decline in Canada (Gidengil and Everitt 2003). Everitt and Camp's (2009a, 2009b) exploration of the depiction of gay and lesbian politicians found that the Canadian media tend to accentuate the "otherness" of homosexuals even when it is of little relevance. Stereotypes and misrepresentations have also been found by Abu-Laban and Trimble (2010), who studied the media's treatment of Muslim Canadians. The prevalence of discriminatory discourses is so systemic and pervasive that Fleras (2011, 4) reasons it "should be cause for concern, a commitment to action, and a catalyst for change."

These trends have several democratic implications. Media attention to political strategy tends to occur at the expense of reporting on political issues. The prevalence of the strategic frame may leave some voters politically illiterate (Gidengil et al. 2004, 66-67). Political parties target their communications to segments of the electorate, and campaign learning is confined to the well informed. Framing may influence voters by encouraging them "to link their vote decisions more closely to their ratings of leaders, and less closely to their evaluation of parties and issues" (Mendelsohn 1993, 165). Unequal and negative coverage of parties calls into question the fairness of reporting. Finally, slanted coverage offers one explanation for why qualified minorities and women might avoid public life. Moreover, it might affect citizens' evaluations of the viability and credibility of such candidates (Norris 1997a).

POLITICAL COMMUNICATION AND CANADIAN CITIZENS
Citizens are the third group of actors that influence, and are influenced by,

the political communication process. The traditional way of thinking about citizens is that they are the recipients of information from political institutions. This approach has informed most research, namely that on the effects of communication activities. Yet many citizens are also now producers of political content. The wide dissemination of the Internet, coupled with the recent developments in online social media, has generated significant interest within the research community on how citizens use these tools to express political opinions. We look at both of these angles in turn.

Influenced by remnants of persuasion theories from the 1940s, research on effects of political communication has been a staple of scientific literature, going back to the 1960s. Three core political impacts of the media have been studied: agenda setting, priming, and framing. The first refers to the capacity of the media to shape citizens' political priorities. In a now infamous quote, Bernard Cohen (1963, 13) first summarized agenda setting as the power the press has in telling people what to think about rather than what to think. This proposition expresses how levels of coverage from the news media could attract citizens' attention to certain political issues and not to others (see also McCombs and Shaw 1972; Iyengar and Kinder 1987). In Canada, Soroka (2002) applied an integrated agenda-setting framework to better understand how the media's agenda, the policy agenda, and the public's agenda interact. A key conclusion was that although the media often set the public agenda, in some cases it is the public that affects the media.

The priming effect of the media is presented as a consequence of agenda-setting effects. Iyengar and Kinder (1987) demonstrated how the news media can shape audiences' political priorities and how a single issue that dominates the media can become the determining factor in voters' evaluation of candidates. In the Canadian context, Mendelsohn (1994, 1996) discovered that high exposure to news coverage of the 1988 federal election primed leadership and leaders' trustworthiness as determining factors for Canadian voters, with partisanship becoming less important in voting decisions. These conclusions were upheld by Gidengil and her collaborators (2002), who found that higher exposure to news coverage increased the impact of leadership in the 1993 and 1997 federal elections and, in 1988, of the free trade issue, which completely dominated the news agenda.

The third media effect presented in the political communication literature is framing. Iyengar (1991) demonstrates how frames of political issues will bring viewers to cast different forms of responsibility on policy makers. Thematic frames usually bring audiences to place some form of responsibility on governments, parties, and political actors. Conversely, episodic coverage is said to limit the capacity of citizens to attach political responsibility to the covered issue. Similarly, other research has indicated that positive and negative coverage of political parties during elections may be associated with movements in voting intentions. In their analysis of the electoral impact of media exposure on the 1997 federal elections, Dobrzynska and her collaborators (2003) found that positive coverage had a political benefit on vote intentions during the campaign, but that the effect did not last until Election Day. Other research has found that framing may affect leaders' evaluations as well (Cappella and Jamieson 1997, 84-85), especially female politicians, who are said to suffer from the dominant strategic framing in electoral news (see, for example, Norris 1997a, 160; Trimble 2005; Bashevkin 2009b, 29; Goodyear-Grant 2009, 161). Thus, when the media frames campaigns, politicians, and issues in certain ways, they provide political cues, social definitions, and information for citizens to make sense of politics and make electoral decisions.

Another considerable portion of the literature dedicated to effects investigates the impact of political communication on political participation and the vitality of democratic life. The media malaise theory argues that contemporary political coverage in news media offers a truncated and negative view of the political process and stimulates citizens' political cynicism, declining trust, and engagement. Despite its popularity, this theory has many opponents who believe a clear empirical demonstration of the association between media exposure and democratic malaise has yet to be provided (Newton 1999; Norris 2000; de Vreese 2005; de Vreese and Elenbaas 2008; Adriaansen, Van Praag, and de Vreese 2010). In Canada, Nadeau and Giasson's (2005) meta-analysis of the media malaise phenomenon came to the conclusion that current Canadian research had not yet presented a strong enough demonstration of the validity of the hypothesis.

A more recent area of research explores how Internet use modifies the communication process. As citizens become active producers of political information, political institutions and the media are forced to react more efficiently to public demands expressed online. One area of scholarship looks at new forms of online participation where citizens use the Internet to engage, debate, mobilize, and act politically (Mossberger, Tolbert, and McNeal 2008; Coleman and Blumler 2009; Pole 2010; Giasson, Raynauld, and Darisse 2011). Scholars surveying political Internet users have identified five broad categories of actions, namely information gathering, use of e-government services, the discussion of politics online, conventional participation (donating to parties, emailing a representative, joining a party, voting online), and non-conventional participation (boycotts, online protests, activism for social movements). For instance, because of several recent, high-profile grassroots movements, such as Occupy, the Arab Spring, and #IdleNOMore, there is a growing body of literature exploring the role that digital technologies are playing in such grassroots mobilization (see Gerbaudo 2012). Another area of study focuses on citizens' information production process, especially in relation to grassroots journalism (Gillmor 2006; Nip 2006). Many political bloggers consider themselves as independent journalists or claim that they blog as a way to monitor or counterbalance biases in the political coverage of conventional news media (Koop and Jansen 2009; Pole 2010).

Finally, another broad portion of the literature looks at the potential of the Internet to stimulate citizenship and democracy in postindustrial societies (e.g., Coleman and Blumler 2009). From its inception, the digital politics literature has hypothesized about a positive relationship between online political activity and democratic citizenship. Cyber-optimists posit what's known as the mobilization hypothesis: that the Internet creates opportunities for the politically disenfranchised and marginalized, including young people and minorities. Cyber-skeptics claim that the digital divide, parties' ideological commitments, and normalizing principles represent constant barriers to real access to online resources and tools. This reinforcement theory states that technologies are used by citizens who are already politically sophisticated, interested, and active offline (R. Davis 1999; Bimber 2003). In Canada, Roy and Power (2012) found limited support in the 2011 federal election for a thesis that the Internet had become a means of mobilizing Canadians who are otherwise politically disengaged. Their data showed some evidence of a reinforcing effect; those most apt to be politically active offline were complementing this activity with online activities. Thus, once again, we are faced with the changing nature of Canadian political communication that is at an intersection of bridging "old" and "new" media.

The Structure of *Political Communication in Canada*

Political Communication in Canada brings together knowledge, viewpoints, and evidence from senior scholars, as well as from new voices in Canadian political communication research. An empirically driven book, featuring innovative case studies, it draws on various methodological approaches and theoretical perspectives. Following a common framework, each chapter is guided by three overarching questions:

1   What tactics, tools, or channels are used by political institutions, by the media, and by citizens in Canada to disseminate information?
2   To what extent is the new political communication environment resulting in a more informed, engaged, and/or cynical citizenry in Canada?
3   What are the corresponding implications of the new political communication environment for Canadian democracy?

Taken together, these chapters provide an assessment of the full range of political communication activities used in Canada, while taking into account the new stage of political communication. They are organized within sections that refer to a component of the political communication triangular process. Part 1 focuses on political communication by Canadian political institutions. Here, chapters are dedicated to the activities of governments, political parties, and interest groups in their attempt to inform, promote, and persuade Canadians. Older techniques such as advertising and media management, and new ones like political branding and social media, are examined. Part 2 takes a look at the Canadian political news media. The chapters explore media from the perspectives of how politics is communicated and how political journalism is practised. Part 3 turns to political communication and Canadian citizens. Consistent with the aforementioned angles of research, the chapters focus on media effects and new forms of online participation by citizens, as well as on political communication from advocacy groups.

Understanding political communication is also important for understanding democracy. Canada, as in other advanced industrial countries, is said to be facing a crisis of democracy, as evidenced by fluctuating participation and interest in politics. Turnout in the 2008 election (58.8 percent of registered voters) was the lowest recorded in Canadian history, and the downward trend was only slightly abated in the 2011 election. There has been a corresponding long-term decline in feelings of efficacy among citizens (Teixeira 1987; Koop 2012). Levels of internal efficacy (citizens' feelings that they can change politics) has remained steady, but levels of external efficacy (citizens' feelings that institutions are responsive to them) have declined. Finally, many citizens also report a lack of interest in politics, despite the increasing availability of political information (e.g., Gidengil et al. 2004). The implication is that uninterested citizens are not typically seeking political information, which in turn might produce information, participation, and interest representation gaps within the citizenry. Thus, the most informed citizens take part in the political system and have their interests represented by institutions, but the values, needs, and aspirations of the least interested, least informed, and least engaged may not be heeded by decision makers.

Political Marketing in Canada provided insights on the degree to which political elites' use of opinion research can hinder and help democracy. In *Political Communication in Canada*, we build on this knowledge by exploring the communication activities of political institutions, the mass media, and citizens. To what extent do political institutions and the news media encourage or discourage citizen participation? Do they provide citizens with enough and suitable information by which to make political decisions? To what extent are digital technologies providing new venues for citizens to engage with political institutions and the media? Does the evolving Canadian communication environment allow for novel types of political action for citizens? These are among the questions raised in this book as we seek to understand the implications of political communication for Canadian democracy in a transformed communication environment.

NOTES

1   The eight media corporations were Astral, Bell Media, Postmedia Network, Quebecor Media, Rogers Communications, Shaw Communications, Telus Communications, and Torstar.

2

The Governing Party and the Permanent Campaign

Anna Esselment

The concept of the permanent campaign embraces the notion that government can and should be run as a continuous campaign. There are no longer writ periods or an election season; instead, the party in government operates as if it were actually fighting an election by, for example, relying on political pollsters and strategists for advice in government, using polls to gauge the public's opinion on policy to build and maintain support, tightly controlling the "message" of government, and oversimplifying issues (often into "for" and "against" positions) as part of the constant battle to win over a shrinking pool of voters (Sabato 1981; Bowman 2000; Heclo 2000; Hess 2000; Needham 2005, 344). Consequently, every action, decision, and communication by government has been strategized, tested, and deliberately conveyed according to an overall theme or message designed to win public approval. The permanent campaign has implications for political communication, since it is essentially the conflation of campaigning and communications into permanent election communications. Governments must be able to consistently and effectively express *what* they are doing and *why* in a format that is both easily digestible by voters and appealing to core supporters. Permanent election communications in government has its roots in American politics. The general framework for the permanent campaign has been constructed through the evolution of the presidency, the partisan polarization of Congress, the dominance of the scientific poll, and new communication technologies. The model has permeated other jurisdictions, including Canada. Much like the centralization of power into the executive branch (Savoie 1999a), it is likely that the permanent campaign will remain an enduring feature of government in Canada.

The permanent campaign evolved within the American political system, where the president is a key figure. Scholars have argued that the president's predominant power lies in the ability to use rhetoric, to bargain, and to persuade in order to pursue an agenda with representatives in Congress (Neustadt 1960; Tulis 1987). Because of the necessary inter-dependence of the executive and legislative realms, the leaders of both branches had to find ways to convince and cajole each other to ensure cooperation on policy matters. Until recently, working with one another through bargaining is in large part how most policy matters have been settled in American politics. As technology has evolved, a more aggressive approach to persuasion is to "go public" (Kernell 1997). Kernell defines going public as "a strategy whereby a president promotes himself and his politics in Washington by appealing directly to the American public for sup-port" (ibid. 1997, 2). In short, the president goes over the heads of congres-sional members and takes issues to the members' constituents. This can be achieved through TV and radio addresses, through travel, and now also through social media channels (Cook 2002; Doherty 2007).

In part, the going public strategy of presidents arose from the increased polarization between the two major parties in Congress, a second compon-ent contributing to the permanent campaign. Congressional studies provide empirical evidence that partisanship has sharpened over the last twenty-five years (Aldrich and Rohde 2000; Jacobson 2007, 23), and Democrats and Republicans are less likely to be swayed into supporting legislation emanat-ing from the other party or from the White House when government is div-ided. With less room for direct bargaining and mutual accommodation, presidents attempt to move public opinion to drive votes their way in the legislature. The strategy necessarily results in a continuous communications campaign to win the hearts and minds of Americans on an issue-by-issue basis that, theoretically, will influence their representatives, who will in turn respond to the demands of their constituents.

The development of the permanent campaign can also be attributed to analyses of public opinion and voting. Conducting market research to un-cover what the public thinks of an issue provides valuable information to a government; that insight will inform its political communications on the subject, especially when it comes to wooing that 40 percent of voting-age citizens who lack a strong party identification (known as swing voters) (American National Election Study 2008). That an administration can gauge citizens' views on a subject, target those whose views align with the president's or those who can be persuaded to align their views, and then craft messages that will galvanize their support secures for public opinion and polling a prominent place in the permanent campaign (Tenpas and McCann 2007). As Sidney Blumenthal (1980, 8) presciently pointed out, however, swing voters are by definition uncommitted – they do not provide dependable support. Consequently, presidents *must* continuously campaign on each issue in order to draw together endless varieties of coalitions; as a result, campaigning and governing become one.

The study of political communication is the last component in the development of the permanent campaign. The concept is underpinned by the emphasis on communications control, carefully crafted messages, and message discipline by members within the executive branch (Faucheux 2003; Burton and Shea 2010; D.W. Johnson 2011; Taras and Waddell 2012b). The campaign-in-government is fuelled by the what, where, when, and how of delivering deliberate and concise missives to a range of audiences in order to capture popular support. Franklin D. Roosevelt's fireside chats on the radio, heard by Americans in their homes, were an early manifestation of the permanent campaign. Decades later, the advent of twenty-four-hour news channels and, later, Web 2.0 fuelled the prominence of permanent campaign communications. The first all-day news channel in the United States was launched by CNN in 1980, and the successful format was eventually copied by other networks. On the one hand, all-day news channels compelled politicians to better manage what they were saying and how they said it in order to avoid missteps that could become events in themselves – a device often used by outlets to fill the news cycle. On the other hand, with a plethora of political programs embedded into news channels, politicians and their spokespeople also seized opportunities to appear on these shows to transmit their message to a large public, often using speaking points to ensure consistency (Burton and Shea 2010; D.W. Johnson 2011).

Beyond the traditional methods of communicating politically (speeches, travel, public events, television adverts, direct mail), political leaders are now also harnessing the power of social media to interact directly with voters. Through web pages, blogs, Facebook profiles, YouTube uploads, Flickr, tweets, and other online presences, members of the government have been given a new tool with which to campaign while in power. The web facilitates deliberate and continuous political communication with core voters (often called a party's base) and potential supporters by enabling the delivery of unfiltered messages directly to, for example, email inboxes and social media accounts (Small 2010b; see also Chapter 6 in this book). With the ability to wield more control over political news – and target intended recipients of those missives – political leaders and their strategists are better able to manage it (Lilleker 2006). This is demonstrated by strategies such as lines of the day that focus news reporting of the government on a certain policy or position it wants to emphasize (Tenpas 2000, 124-25). Similarly, political photography – or photo of the day – is a common strategy political leaders and their advisers use to assert control over the government's image and sell its policies (Davies and Mian 2010; Marland 2012; see also Chapter 4 in this book). With a greater capacity to set the news agenda and disseminate messages deliberately targeted to shore up policy support, governments are better positioned in the fight to win public approval for their policies. Doing so on a continuing basis, however, requires that campaign-like strategies and tactics be imported and executed while in office.

What does this all mean? The credibility and legitimacy of politicians and their key issues are now inextricably linked to positive public support. The line between campaigning to win government and governing itself is increasingly blurred. In short, everything becomes a campaign to win public approval, and communication tools are employed to secure that crucial support.

Most scholars find the rise of the permanent campaign troubling. With its emphasis on message and communications control, academics fear that political leaders are manipulating citizens on a wide scale. Blumenthal (1980, 8) lamented that "the citizenry is viewed as a mass of fluid voters who can be appeased by appearances, occasional drama, and clever rhetoric." Heclo (2000, 33) echoed this sentiment, noting that "the permanent campaign is a school of democracy, and what it teaches is that nothing is what it seems, everything said is a ploy to sucker the listener, and truth is what one can be persuaded to believe." In other words, issues are simplified, yet the substance of these policy debates is not simple. The assumption that citizens must be fed sound bites and carefully prepared images because they cannot or will not assess policies with reasonable thoughtfulness is presumptuous of those in control of government. The permanent campaign, observed Heclo (2000, 29), "is not the way Americans do politics, but the way politics is done to them."

As mentioned, the permanent campaign is not exclusive to politics in the United States; signs of it in other jurisdictions are also apparent. Despite its denials to the contrary (Gould 2002), Britain's New Labour government under Tony Blair (1997-2007) has been criticized for its permanent campaign, particularly for its branding and communications strategies

(Needham 2005; Scammell 2007). Similar analyses have been made in Australia (Van Onselen and Errington 2007; MacDermott 2008), Italy (Roncarolo 2005), and Ecuador (Conaghan and de la Torre 2008). Very little literature exists about the permanent campaign in Canada and thus this is an underinvestigated area of Canadian politics. There are general accounts of press relations with prime ministers (Gossage 1986; Levine 1993). Savoie (1999a, 2010), along with L. Martin (2003) and Simpson (2001), have focused scholarly attention on the concentration of power in the Office of the Prime Minister. Flanagan (2012) and Rose (2012) see the increasing amount of partisan advertising occurring between elections in Canada as evidence of a permanent campaign, although their examination is focused more on the activities by the extra-parliamentary party organization. Marland (2012; also Chapter 4) has noted the penchant of the governing Conservatives to provide media outlets with staged photographs of the prime minister and various cabinet ministers in its efforts to assert control over the government's public image. The following case study scrutinizes the Conservatives under Stephen Harper and how they adhere to the main features of the permanent campaign framework. Particularly in the way it has used communications tools in government, the permanent campaign, like its counterparts in other jurisdictions, has permeated Canada's government.

Case Study

The Canadian prime minister operates in the same media-concentrated, adversarial environment as most other heads of government. With Canadians spending an increasing amount of time in front of the computer to email, network socially, and find information, the government must use similar communications strategies to reach both core and potential supporters in its quest to convey a positive image and win approval for its policies. When Stephen Harper and the Conservative Party formed a minority government in 2006, message discipline was expected, as the Conservatives would have to run a tight ship to stay in power. Although the party had done well at improving its brand, advertising, and messaging between the 2004 and 2006 campaigns (Flanagan 2007), few imagined the strict command-and-control communications strategy that was swiftly installed, though it was reflective of the permanent campaign mentality. As with congressional members who must be re-elected every two years, the average length of a minority Parliament in Canada is just under two years (Russell 2008). The Harper government *was* still in campaign mode and determined to manage the news by

FIGURE 2.1

Features of the permanent campaign in Canada

- Use of campaign strategists in government for communications advice.
- Command-and-control-style message discipline.
- Reliance on market research and polling.
- A communications style that reflects sharp partisanship.

immediately setting the agenda and framing issues its way. As a result, four distinct features of the permanent campaign model can be identified in the modus operandi of the Conservative government: (1) use of campaign strategists for political communications advice in government, (2) strict control over communications to reinforce the preferred message of government, (3) reliance on polls to help guide communications strategy, and (4) sharp partisan divisions that affected how the government communicated with voters (Figure 2.1).

First, in line with the observations of, among others, Dulio (2004), Needham (2005), and Tenpas (1997), Stephen Harper installed several key campaign strategists into positions of power in the Prime Minister's Office (PMO). Notable people included Ian Brodie (formerly the executive director of the Conservative Party and a campaign manager in 2004 and 2006) as chief of staff; William Stairs, Kory Teneycke, Sandra Buckler, and Dimitri Soudas as communications directors; and Bruce Carson, Keith Beardsley, and John Weissenberger as key advisers who were also part of the campaign inner circle. Hiring partisans to advise the prime minister is not new in Westminster systems, since many prime ministers have kept trusted aides at their side throughout their tenures as heads of government; what is evolving is their sheer number in certain departments within the PMO, highlighting the importance of those particular "shops." Since 2006, the PMO has had between 85 and 110 staffers, with the current number at about 85 (Canada 2014).[1] In spite of scarce information on partisan advisers in Canada, we know the staff are relatively young and very ideological, and have little experience in government (Benoit 2006, 159). They *do* have knowledge or experience in elections and communications, however, and these talents are brought to their jobs in the PMO. Consequently, a campaign-like atmosphere inevitably envelops the PMO as political staffers continue their work to develop and sell government policies. If only for their size, two PMO branches deserve special attention: communications, and tour and scheduling. The communications office has 19 staffers, and 15 work in tour and scheduling (Canada 2014). By comparison, the policy shop has 8 staffers, as does issues management. If we examine these departments in 2001 under former prime minister Jean Chrétien, we find that only 13 people handled communications, and just 6 handled the PM's tour schedule (Government of Canada 2001). During Paul Martin's tenure as prime minister, about 10 staffers were charged with responsibility for communications and 8 for tour (Government of Canada 2006). Interestingly, the Chrétien, Martin, and Harper PMOs retained approximately 8 political staff who concentrated on policy (Government of Canada 2001, 2006).

In the past nine years (2006-14), there was almost a twofold increase in the number of political staff handling the PMO's communications with other departments and branches of government, and with the public. A similar trend was evident in tour and scheduling. Communications and tour are both critical to the permanent campaign: one for message discipline and press relations, and the other to plan the political travel of the prime minister, as strategic visits to certain parts of the country can shore up political support. The swelling of these departments in the Harper PMO highlights the emphasis placed on good political communications by this governing party. It also serves as a reminder of the growing role and influence of campaign strategists at the centre of government.

A second feature of the permanent campaign model now apparent in the PMO is tight control over news management generally and the Ottawa press gallery specifically. As Newman (1999, 112) noted in his book about the mass marketing of politics, "The interest level in politics and the White House drops off significantly after the campaign is over, leaving tremendous power in the hands of the media to shape the president's image through the snippets of information shown on the evening news each night." The Harper PMO was determined to curb the media's power, especially since it harboured the view that many in the media were overly sympathetic to the federal Liberals and naturally conformed to a liberal bias; the PMO assumed that prominent members of the media would be unwilling to portray any of the government's policies in a positive light (Flanagan 2007, 283; L. Martin 2010, 22). Control from the centre was imposed in four ways. First was the creation of the Message Event Proposal (MEP), a system employed by the PMO to vet communications "products" such as news releases, speeches, and public statements of ministers, government backbenchers, and public servants (Campion-Smith 2008; L. Martin 2010, 58). The MEP had been successful in keeping Harper and the Tories on message when they were in opposition; it made even more sense to continue the MEP in government to ensure message discipline and prevent embarrassments should a caucus member, minister, or public servant muse publicly about issues the government preferred not to discuss. The PMO's command-and-control approach was viewed as problematic by some, particularly as the MEP system had the effect of "muzzling" civil servants in a way they had never been before.[2] For the PMO, the MEP was key to running an orderly and effective minority government – the fewer communications mistakes, the better.

A second method of communications control was to modify the way the prime minister dealt with the Ottawa press gallery (see Chapter 7 in this book). Uncomfortable with the usual method of press gallery relations where the PM and ministers would face a barrage of reporters' questions as they left cabinet meetings, attended news conferences, or visited with dignitaries, the Harper PMO imposed immediate changes on taking office. Among other things, the date and time of cabinet meetings were kept secret to protect ministers from the onslaught of media questions about their portfolios; journalists wishing to question the PM at a media availability were required to put their names on a list from which PMO staff would select who would be permitted to pose their questions (L. Martin 2010, 65); and in-house photos would be shot and provided to the media by the PMO when dignitaries visited or when a certain visual was important to the government's message (Canadian Press 2011a; Marland 2012). Even planned events for the prime minister and key ministers to make government announcements were tightly controlled, allowing only partisan supporters to attend (L. Martin 2010, 130-31). These measures scaled back "meet the press" opportunities but increased the ability of the PMO to influence the type of image it wanted to portray to the public. The prime minister was available to journalists, but his office selected the questioner. Providing carefully managed photos to the media meant the visual image would reinforce how the government wanted to be perceived by the public. Finally, stacking news events with core supporters ensured that only a positive response to a government announcement would be reflected to other Canadians paying attention to news coverage.

A third way to impose communications control was to permit only a handful of ministers (in addition to the communications director to the PMO and the prime minister himself) as spokespersons for the government. When elected in 2006, many of the MPs in the Conservative government were new to Parliament Hill. A few of the members with more polarizing right-wing views had been tight-lipped during the campaign, and the PMO wanted to keep these ideologues quiet in government. Unnecessarily sparking debate on emotional issues would be unhelpful to a government hoping to achieve majority status in the next election. Consequently, Canadians heard primarily from Ministers John Baird, Jim Flaherty, Jason Kenney, Peter MacKay, Lawrence Cannon, Tony Clement, and Maxime Bernier. Each was well spoken and rarely strayed from the government's message for that day or particular issue. Most had been elected politicians for ten years or more, and a few had experience as ministers in provincial governments. All were fiercely partisan and loyal to Stephen Harper. Given the intense media scrutiny under which it operates, it was understandable for the PMO to curb the number of spokespeople. At the same time, it was different for Canadians to hear only from a handful of politicians rather than from the various ministers of the Crown. The imperative of the permanent campaign, however, is to seek and maintain public approval, a feat usually involving those most adept at spinning government policy into the best possible light.

Finally, controlling communications also meant that the PMO would carefully monitor the media itself and, much to their surprise, its own MPs. In documents tabled to the House of Commons, the Harper government spent $23 million between 2011 and 2013 to track media accounts of government programs and services, hot-button issues, what Conservative MPs were saying, and public perceptions of its ministers (Cheadle and Levitz 2012; *Huffington Post BC* 2013). The extent of the tracking and the amount of funding devoted to it suggest that the government has a keen interest in knowing what Canadians are thinking and saying about its performance. This type of media oversight fits with the model of the permanent campaign: where MPs go off message and the media reports it, the PMO can bring them back in line; where ministers are not being perceived well by targeted voting segments, course corrections can be made to change how these groups view the government. That most of the Conservative MPs were surprised to discover they were being tracked by the PMO suggests that the Office of the Prime Minister remains the locus of centralized power, particularly as the permanent campaign becomes a more prominent part of governing in Canada.

A third feature of the permanent campaign model evident in the Canadian context is the need to gauge and reflect public opinion while in government. Governments may engage in polling and other public opinion measures to test reactions to new policies, to see how Canadians perceive government services, to assess government approval ratings, to identify key public issues, and to collect valuable information on how to frame issues and guide a communications strategy (Bowman 2000; Tenpas and McCann 2007). As part of its permanent communications machinery, the Harper government partook in extensive polling in its first year of government. The annual cost was $31.4 million – the most any government had spent on public opinion research (Public Works and Government Services Canada 2008). Forty-four percent of this total was spent on marketing and communications research and on advertising initiatives, which accords with the permanent campaign framework. Another 40 percent was spent on policy development and program evaluation, with the remainder spread among the categories of quality of service, web testing, and "other" (ibid.).

The use of taxpayer dollars to fund the government's appetite for polling did not go over well with the media. Substantial market research initiatives fell off after the first year, with only $6.5 million spent on public opinion research in 2011-12 (Public Works and Government Services Canada 2012). This significant drop in government polling grates against the expected behaviour of a government engaged in constant electioneering but suggests that there may be some expenditure limits for which the public is willing to pay for government market research. Some observers suggest the government has not halted public polling but instead shifted the bulk of its cost to the Conservative Party, where the results can be used more overtly for electoral purposes (L. Martin 2010, 119). Considering the fundraising largess of the Conservative Party of Canada, this is a distinct possibility. Of note, however, is that an "Advertising and Market Research Unit" was installed in the PMO in 2009. This suggests that the government remains interested in what the public thinks and, we expect, how it can most effectively sell key messages to Canadians.

Hyperpartisanship is the fourth feature of the permanent campaign in Canada, and this is also found in the Harper PMO: there is a sharp partisan division between the government and the opposition parties, which often results in a communication strategy that emphasizes negative politics. The prime minister himself has been criticized for displaying overt partisanship (Johnson 2005), and members of Harper's front bench, such as John Baird, Tony Clement, and Peter Van Loan, as well as many staff in the PMO, are also extreme partisans. Now parliamentary government is party government, so partisanship is to be expected. The hyperpartisanship can be seen most obviously in the practice of wedge politics and attacks on the opposition. Wedge politics can be divisive and devastating. Complex policy problems are usually framed two-dimensionally and deliberately oversimplified for a straightforward political message that aims to strengthen a party's base of support while also potentially stripping one's opponent of their adherents (Ward 2002). This has occurred most often with the war in Iraq, support for Israel, and the issues of crime and gun control (L. Martin 2010, 126; Giasson and Dumouchel 2012). If an opposition party votes against a bill that increases defence spending, for example, it may be accused of supporting terrorists and putting Canadian troops in danger. A proposal to put more funding into rehabilitating offenders can be spun to suggest a party is soft on crime. These messages appear to resonate with the public because of their simplicity and overtones of negativity.

Using government resources to communicate a position on wedge issues is common in the permanent campaign (Van Onselen and Errington 2007). MPs, for instance, are granted franking privileges, which means they can send newsletters to their constituents without paying postal fees. These householders often ask the constituent for feedback on a certain policy matter, frequently by giving a choice between policy positions held by the government and the Opposition, providing the governing party with additional public opinion research. Although now banned, the use of out-of-riding "ten percenters" – where MPs could send householders equivalent to 10 percent of their own constituency to ridings other than their own – was a communication tool intended for voters in ridings held by opposition parties. For the governing Conservatives, the message and geographic target of the householders were often coordinated by the PMO for political advantage (*Ottawa Citizen* 2008). The overtly partisan content of these publicly funded direct mailings in recent years led to a ruling by the House of Commons Board of Internal Economy to end out-of-riding ten percenters (Parliament of Canada 2011).

The use of government-funded partisan advertising for communication is not limited to wedge politics. Broadly defined, a partisan communication promotes the partisan interests of the governing party (Office of the Auditor General of Ontario 2012b). Conservative MPs have been caught in photo ops handing over ceremonial government cheques with the Conservative Party logo broadly displayed (CBC News 2009). Prime Minister Harper has used his Commons office as the backdrop for a partisan commercial aimed at Tory supporters (Naumetz 2011). Bureaucrats have been told to replace the words "Government of Canada" with "Harper Government" in their ministries' news releases and background documents (Canadian Press 2011a). In the winter of 2011, a massive outlay of public dollars – $26 million – was used to advertise the government's economic action plan (Canadian Press 2011b). Commercials that provide information to Canadians about government services are common practice; ads that suggest support for the government based on their policies (as contained in a number of the economic action plan commercials) are more partisan in nature (see Kozolanka 2012). Using public funds for partisan ends has been practised by other governments in Canada; in fact, because of the extensive use of partisan advertising by the Progressive Conservative government in Ontario (1995-2003), this particular use of taxpayer dollars is now prohibited in that province (Ontario 2004). There is, however, no equivalent law at the federal level. Consequently, publicly funded partisan communications to Canadians is becoming an entrenched part of a continuous cycle of campaigning.

Sharp partisanship is also evident in the way the Conservatives attack their opponents. The PMO has orchestrated a few of these attacks, most recently by leaking unsolicited information to the press about current Liberal leader Justin Trudeau's speaking fees (Watt 2013), and by giving cabinet ministers props to bring into the House of Commons to mock Opposition leader Thomas Mulcair (CTV News 2013). Other blasts at their opponents occur through the extra-parliamentary party organization via traditional communication tools (direct mail, television advertising) but are increasingly being moved to online resources (Small 2012). Stéphane Dion was Liberal leader for only two months before commercial offensives were launched, branding him weak and hapless. Within weeks, Dion's approval rating was at a dismal 17 percent – versus Harper's 42 percent (Nanos 2007). Michael Ignatieff, the next Liberal leader, faced a similar fate when Conservative inter-election commercials depicted him as "just visiting" from the United States and Britain, where he had spent much of his time as a professor at Harvard and Cambridge. Websites were also set up to "e-ttack" the Liberals and NDP (Small 2012). The Liberals were hard-pressed to respond, and the party lost so many seats in the 2011 federal election that it became, for the first time, the third party in the Commons. Both Flanagan (2012) and Rose (2012) note that political party-funded campaign advertising between elections is on the rise and, because of its fundraising prowess in recent years, the most pre- and post-writ advertising has been by the Conservative Party.[3] In fact, it is here we see the interconnection between political communication and fundraising. Effective political communication to supporters through both traditional and new communications technology has led to an influx of funds for the governing party. In turn, the party has used that money to communicate its message more broadly. The Conservatives have effectively eviscerated two Liberal leaders, torn apart Opposition party policy (such as the Liberal Green Shift), ended any chance of a coalition government in 2008, and increased support for certain legislative initiatives (Flanagan 2012, 129-36).

Despite winning a majority government in 2011, the permanent campaign cycles along. The partisanship is still sharp and divisive, and attacks on the opposition continue apace (Wherry 2012). The government keeps a tight rein on communications, whether it is about responding to media inquiries (Spears 2012) or controlling its image to the public (Marland 2012). This suggests that the nature of the government (either minority or majority) does not so much determine permanent campaign practices as does the type of communications environment in which the government must function. Where PMO adviser Keith Beardsley thought the "clampdown" would be short-lived after coming to office, it instead became the dominant modus operandi of government (L. Martin 2010, 27).

Political Communication in Canada

It can be argued that Harper's command-and-control style is not so different from that of Liberal prime minister Jean Chrétien. Chrétien's time in power led to the scholarly observation that power had been concentrated in Canadian politics (Savoie 1999a). It was Chrétien, not Harper, who was dubbed a "friendly dictator" (Simpson 2001). Did the permanent campaign originate in Chrétien's PMO? Analyses thus far suggest that continuous campaigning has its roots with the Harper PMO (Kozolanka 2012, 117-18). It is far more controlling than Chrétien's, and for various reasons. For one, Chrétien led three majority governments and was not preoccupied with the possibility of an election should his government fall.[4] Chrétien also had extensive experience at the federal level, serving in the governments of Lester Pearson and Pierre Trudeau. He had more trust and confidence in the public service. Chrétien's government further differed from Harper's in that it had a deep pool of talent on the front benches – cabinet ministers were generally quite capable and left to handle their portfolios on their own, including the decision to talk with the press gallery (Goldenberg 2006, 73-75). Significantly, Chrétien's PMO functioned only on the margins of Web 2.0. Email and online communication were prominent, but Facebook launched the year *after* he left office. Other social networking sites (such as YouTube, Twitter, and Flickr) were founded shortly thereafter. The reduced size of his communications and tour shops within the PMO – as compared with Harper's – is also evidence that the permanent campaign had a lesser presence during Chrétien's tenure as prime minister.

That the permanent campaign has appeared to settle over Ottawa – as it has in other jurisdictions, including the United States, Britain, and Australia – is probably of little surprise, but perhaps it will be of consequence. The main debates outlined earlier remain: in spite of Key's (1966) argument fifty years ago, are voters now thought to be fools? Will they accept packaged images, a moribund press, and spoon-fed sound bites? Will they become complacent in their desire to critically assess policy choices? Is the permanent campaign and its obsession with polls, political spin, and partisanship contributing to the tumbling rates of voter turnout?

The answer is still unfolding, though two things are clear. Whether viewed as good or bad, the permanent campaign is here to stay. Savoie (1999a) argued that the centralization of power in Canadian politics was institutionalized; in other words, it was not a function of individual personalities – parliamentary government had in effect become prime ministerial government. The same will be true of the permanent campaign model; in a communications-saturated environment, it too will become an institutionalized feature of our political system. Individual personalities will, however, matter here. It is not that one prime minister will halt the campaign; instead, personalities will have an effect on the intensity with which the permanent campaign is waged.

The political environment that remains is both better and worse. It is better in the sense that the desire for public approval by governments is not a lamentable outcome; public approval is given when governments are responsive, and responsive government can also be good government – the two need not be exclusive. The political environment is worse in the sense that, among other things, the permanent campaign model often sets aside cooperation, negotiation, and moderation among the political players. It can dampen morale in the public service and impose new limits on the role of the MP, and it also requires sidestepping the fourth estate in order to impose professionalism and control over image making. Further, the communication tools inherent to the permanent campaign demand the concise relaying of information, usually to the detriment of policy substance and nuance. And finally, the permanent campaign and its inherent negativity can create cynicism among voters. Those who become alienated by it are removing themselves from political participation, presenting a significant challenge to Canadian democracy.

NOTES

1   The number of partisan staff in the PMO swelled under Prime Minister Pierre Trudeau, and the average since then has been about ninety. This can be compared with the number of political advisers in the British PMO, which in 2004 stood at twenty-seven (see Blick 2004, 256).

2   The auditor general flatly refused to have any of her press releases or public statements vetted by the PMO, as this would be a breach of the independence of parliamentary officers (see Campion-Smith 2008).

3   The Conservative Party spent over $550,000 on political advertising in 2010, as opposed to $359,000 for the NDP, and $231,000 for the Liberals (see Rose 2012, 160). For more on advertising, see Chapter 3 in this book.

4   From 2003-6, Paul Martin was Liberal prime minister in a minority government. Despite using political communications strategies to try to offset damage from the sponsorship scandal, his office did not employ the full range of available communication and marketing tools (see Kozolanka 2012). The tenure of Stephen Harper's government is thus the beginning of the true permanent campaign in Canada.

3
Cognitive Effects of Televised Political Advertising in Canada

Pénélope Daignault

The preceding chapter's brief discussion of political advertising invites further analysis. Television political ads are a critical component of any modern election campaign strategy and a foundation of political marketing around the world (Kaid 2012). Candidates and political parties spend a considerable amount of their overall campaign budget on advertising in order to convey major themes, make comments about the opposition, and discuss each other's personal qualities (West 2009). There is, accordingly, a vast body of work in political communication focused on the nature and impact of television political ads. This is particularly true for negative ads, where existing work tracks the trend toward negative advertising in the United States, as well as the impact of those ads on, for instance, the size of the electorate (e.g., Ansolabehere and Iyengar 1995; Fridkin and Kenney 2001; Nesbitt-Larking 2009), voters' cynicism (e.g., Kaid et al. 2007; Jin, An, and Simon 2009), and persuasive processes, including attentiveness and memorization (e.g., Bradley, Angelini, and Lee 2007; Cheng and Riffe 2008). There is, however, very little work from a communication perspective exploring the impact of television political advertising outside the United States.

Researchers usually differentiate between two types of televised political ads, positive and negative, the latter also called "attack ads" and sometimes subdivided into image ads (focusing on negative personal attributes of a candidate) and issue ads (focusing on particular candidates' stances on important issues or policies) (Ansolabehere and Iyengar 1995; Dardis, Shen, and Edwards 2008). Other researchers add a third type of advertising, the comparative ad, also known as the contrast or mixed ad (e.g., Meirick 2002), which combines negative and positive elements, such as criticizing the opposition and presenting a candidate's own perspective. Thus, whereas a negative ad presents exclusively negative information about a competing candidate for the purpose of imputing inferiority to the target, a comparative ad directly compares two candidates on specific points so as to present one in a more favourable (and one in a more negative) light. In turn, a positive ad is regarded as one that exclusively highlights the positive attributes and realizations of a candidate or party.

Research in media and social psychology suggests that individuals react more intensely to negative information than to positive information. Evidence shows that not only do political negative ads elicit more attention but also they are more easily recalled than positive ads (Bradley, Angelini, and Lee 2007). The effects of negative TV spots on subsequent political attitudes and behaviours are still being debated. Some argue such ads have a positive impact on political participation, for one (e.g., Marcus, Neuman, and MacKuen 2000; Goldstein and Freedman 2002). Others suggest that they can trigger a boomerang effect, meaning that the public's reaction runs counter to the desired impact of the ad, notably by enhancing political cynicism (Dardis, Shen, and Edwards 2008) and ultimately reducing voter turnout rates (Ansolabehere and Iyengar 1995). These last findings are, however, refuted by two meta-analyses in particular, which indicate that there is no relationship between negative campaigns and turnout (see Lau et al. 1999; Lau, Sigelman, and Rovner 2007). Scholars (e.g., Wei and Ven-Hwei 2007; Cheng and Riffe 2008) also highlight a third-person effect – that is, the tendency for some voters to project an influence of negative political ads on to others but not on to themselves.

The tone of political ads, usually qualified as either positive or negative, is often solely determined by the dominant valence of verbal or written argumentation. The dichotomous categorization based on information content is problematic, however, as it obscures differences between various levels of tone and ignores the non-verbal elements of advertising, such as imagery and music, the purposes of which are to increase the credibility and liveliness of the ad and make it more convincing (Brader 2005, 2006). It follows that, alongside the more traditional focus on the text of ads, it may be important to capture their audio-visual cues, or emotional cues, as Brader refers to them.

Emotional cues (e.g., bright or dark colours, laughing or children's cries, lively or disturbing music) can be used to evoke positive emotions such as joy, and negative emotions such as hate or fear. Numerous studies present solid evidence in support of the effect: both these pleasant and unpleasant stimuli produced the expected responses (e.g., Bradley and Lang 1999; Juslin

2001). For example, disturbing music using high-pitch sounds can increase anxiety, whereas harmonious music using lower-pitch tones increases enthusiasm (Juslin 2001).

For an advertisement to have any effect, one has to be exposed to it. Compared with ads solely broadcast on political parties' websites, for instance, television ads are successful in overcoming selective exposure, thus ensuring that voters are exposed to ads from all sides of a campaign (Kaid 2012). The most tangible impacts of political ads are actual voting decisions. However, confirming the direct effect of political advertising on voter electoral choices is a complex undertaking, particularly in a context where voters vigorously pronounce their disapproval of such ads, especially negative ones (Franz et al. 2007). Impacts may be more indirect. Ads can influence what voters know about candidates and public policy and how engaged citizens are in their own governance (Goldstein and Ridout 2004). Ads may provide information to voters as they weigh their voting options; they may shift their view of the campaign and condition the way in which voters interpret or seek out information later in the campaign. In influencing what voters know, ads can influence whom voters choose.

Some research focuses on effects such as voting behaviour; another body of research focuses, typically within an experimental lab, on the short-term attitudinal, emotional, and cognitive responses to ads. This work has been critical to understanding ads as a means of campaign communication. It has helped us to better understand, at the individual level, if political ads matter, and how and why they matter. Initially developed by Greenwald (1968) in the field of social psychology, the cognitive response approach gives prominence to the essential role played by individuals' cognitive activity while exposed to a persuasive message. This approach constitutes a paradigm shift away from the Yale attitude change model instigated by Carl Hovland and his colleagues in the early 1950s, which was based on learning theories. Greenwald was one of the first scholars to consider the receiver as an active information processor. He demonstrated the importance of cognition as a mediator of the impact of a persuasive message, and he argued that the content of cognitive responses may be more essential to persuasion than is the learning of communication content.

In line with this approach, Petty and Cacioppo's Elaboration Likelihood Model (ELM) (1981) introduces the concept of cognitive elaboration, which refers to the mental effort expended while processing a persuasive message. The ELM distinguishes between two routes of persuasion positioned along a cognitive elaboration continuum: (1) the peripheral route, associated with a low mental effort, where the receiver uses the superficial cues (e.g., attractiveness of the source, visual elements) to accept or reject a persuasive message, and (2) the central route, associated with an important mental effort, where a receiver analyzes the arguments to form an attitude. The route an individual chooses in a given persuasive situation is based on a combination of motivation and capacity to process the message, but also on factors such as need for cognition, degree of education, and personal relevance. Choosing the central route, according to Petty and Cacioppo (1981), tends to lead to changes in attitude that are more durable and more resistant to persuasion.

Although the ELM has been a theoretical model of choice in many studies on persuasive communication, it has been criticized for not accounting for secondary – peripheral – attributes that may be processed with a high mental effort, as well as for the possibility that central arguments may be the subject of weak cognitive elaboration (Fourquet and Courbet 2004). Unlike the ELM's creators, I argue that an individual can process all types of information – whether central or peripheral – with variable degrees of cognitive elaboration.

Introduced in work on advertising by P.L. Wright (1973),[1] the cognitive response approach is based on the postulate that the persuasive impact of an ad is determined primarily by the scope and nature of the thoughts generated during the exposure phase, thus reinforcing the relevance of studying short-term effects of ads. Wright argued that certain types of spontaneous cognitive responses reflected the psychological processes underlying persuasion in a way that escapes more structured measurement alternatives. Originally developed by Greenwald, the spontaneous cognitive response method (or thought-listing task) allows the measurement of direction and amplitude of attitude change by asking receivers to verbalize their thoughts as they are exposed to a persuasive message.

Existing categorization systems used to analyze cognitive responses and determine the extent of cognitive elaboration typically rely on a combination of valence (positive versus negative) and quantity of cognitive responses. A few systems also focus on types of responses, allowing a more nuanced evaluation of cognitive processing of ads. Nonetheless, in numerous studies of persuasive communication using the cognitive response approach, the only categorization criterion used to code cognitive responses is valence, where favourable thoughts are coded +1, unfavourable thoughts are coded −1, and neutral thoughts are coded 0. This classification does not take into account the strength of cognitive elaboration, nor does it differentiate between types of cognitive responses. To transcend those limitations and better assess the way voters process political ads, a more complex categorization system of cognitive responses is required, and which I discuss below.

The cognitive response approach has become common in advertising and psychology, but it has seldom been employed in studies of political advertising (Meirick 2002). And yet, the way voters cognitively elaborate on TV spots is likely to influence subsequent political attitudes and behaviours. The existing studies of political advertising based on the cognitive response approach mostly use a simplistic categorization of thoughts. Hence, they do not account for different types of cognitive responses (for more details, see Pentony 1998; Schenck-Hamlin, Procter, and Rumsey 2000).

Meirick (2002) is one of a few scholars to have studied the cognitive processing of political ads by casting a broad net when it comes to the cognitive responses examined. His analysis focuses on Wright's categories, as well as on the categories of *source bolstering* (reflecting trust in a sponsor's or an ad's overall means), *identification of negativity* (statements reflecting awareness that an ad is negative), *positive affect* (expressing pure liking for the ad or the candidate) and *negative affect* (expressing pure dislike for the ad or the candidate). By studying American negative and comparative political ads to see whether they differ in the cognitive responses they provoke, Meirick found that comparative ads were greeted by more support arguments than were negative ads. They also met with more statements of positive affect and more source bolstering than did negative ads.

More recently, Phillips, Urbany, and Reynolds (2008) studied the influence of American voters' prior preferences on cognitive elaboration of negative and positive ads, using the thought-listing procedure. Their categorization was solely based on Wright's original categories, namely *counterargument, support argument,* and *source derogation*. The authors shed light on the interaction between political affiliation and cognitive elaboration, demonstrating that confirmatory ads – that is, ads confirming actual preferences – lead to less counterarguments and more support arguments than do disconfirmatory ads.

To my knowledge, no Canadian studies of political advertising have used the cognitive response approach as a theoretical and methodological framework to examine the extent of voters' elaboration, nor the type of thoughts prompted by different tones of ads. For this reason, I employed such a framework in the case study described below.

Case Study

The aim of this study is to assess the cognitive impact of positive, negative, and comparative televised political ads – the majority of which aired over the course of the 2011 Canadian election campaign (see Table 3.1). The main research questions were: What influence has the tone of the ad on the profile of cognitive responses elicited, and on the extent of receivers' cognitive elaboration? What impact has political affiliation had on electors' information processing? Which type of cues – emotional/peripheral versus verbal/central – do voters process when exposed to political ads?

As mentioned earlier, receivers react more strongly to negative information than to positive information. Regarding this negativity bias and its impact on information processing, a body of research suggests that people devote more cognitive energy to thinking about bad things than about good things (Fiske 1980; Abele 1985). In the specific context of political advertising, these rare studies show that negative ads meet with more cognitive resistance compared with positive ads (Pentony 1998; Schenck-Hamlin, Procter, and Rumsey 2000; Meirick 2002). Based on these findings, I formulate the following hypothesis: *H1: Canadian negative televised political ads are conducive to more thoughts that run counter to the ads' desired impact compared with positive and comparative ads.*

I also draw from the Elaboration Likelihood Model to formulate two other hypotheses. People are exposed daily to an enormous amount of advertising. Therefore, they cannot process all of this information with an equal amount of cognitive effort. In fact, most ads are processed with little cognitive effort, via the peripheral route, where receivers use superficial cues to make up their mind. Several models of persuasion indicate that personal relevance of the message is a prerequisite for persuasion (McGuire 1964, 1989; Petty and Cacioppo 1986; Perloff 2003). In my study, perceived personal relevance would be higher toward an ad sponsored by a political party the individual supports. Moreover, as Phillips, Urbany, and Reynolds (2008) argued, political affiliation is likely to influence how an ad is processed. Their findings show an increase in support arguments and fewer counterarguments toward ads that confirmed participants' prior vote intention. Consequently, I formulate the following hypotheses: *H2: Ads are generally processed with a low cognitive effort, prompting more thoughts related to the peripheral route than to the central route;* and *H3: Ads sponsored by the political party that an individual supports are processed with a higher and more favourable cognitive elaboration than ads sponsored by other parties.*

METHOD

So far, the effects of political advertising have mainly been studied through surveys, which measure, in a very general way, the political attitudes and voting intentions. However, surveys cannot reflect the type of spontaneous reactions that I aim to evaluate. Experiments are better suited to the study of short-term responses in political communication (Glaser and Salovey 1998; Brader 2006). Thus, I chose the experimental method to test my hypotheses.

In total, twenty-four TV spots were chosen for the experiment (see Table 3.1). These ads were selected out of 145 ads, in both French and English, broadcasted during the 2008 and 2011 election campaigns and gathered from the websites or YouTube channels of the four main political parties. Most ads were run nationally, though some ads were intended for Quebec voters only. A content analysis of all 145 ads was performed by two research assistants to determine the tone, based on a combination of rhetorical content (type of arguments and themes used) and emotional cues (i.e., type of music, sound effects, colour tones, visual effects), as Brader (2005, 2006) suggests.

The majority of ads chosen for the experiment (twenty) aired during the 2011 election campaign, except for two spots from the Bloc Québécois (BQ) and two spots from the New Democratic Party (NDP). Since the experimental sessions were conducted early in the last federal campaign (i.e., before the parties had released all of their 2011 ads) and there were not many spots from these two particular parties to choose from at the time of the final selection, I had to widen my corpus to include these four French-language spots from 2008. In the end, the experiment focused on seven ads from the NDP, seven from the Conservative Party of Canada (CPC), seven from the Liberal Party of Canada (LPC), and three from the BQ, with the latter seen only by francophone participants. These twenty-four ads were selected to be representative of the three tones: negative (eight), comparative (eight), and positive (eight). The two official languages are also represented, with twelve in English (four from each of the three parties campaigning throughout Canada) and twelve in French (three from each of the four parties).

Participants were recruited through ads posted around McGill University campus. In all, nineteen women and twelve men – all undergraduate and graduate students from various academic programs – participated in the experiment. Of these participants, thirteen were francophone and eighteen

Table 3.1

Name of ad, party, and tone

Name of ad	Language (EN/FR)	Party	Tone of ad
Canada	EN	CPC	Positive
Human smuggling	EN	CPC	Comparative
Tax hike	EN	CPC	Negative
Ignatieff coalition	EN	CPC	Negative
Fardeau fiscal	FR	CPC	Negative
Parti des régions	FR	CPC	Positive
Faire face au défi	FR	CPC	Comparative
Abuse of power	EN	LPC	Negative
Economy	EN	LPC	Positive
Liberal Family Care Plan	EN	LPC	Positive
Liberal Family Pack	EN	LPC	Positive
Avions	FR	LPC	Negative
Allègements fiscaux	FR	LPC	Negative
Démocratie	FR	LPC	Positive
Home heating	EN	NDP	Comparative
Ottawa stopped working?	EN	NDP	Comparative
Where do we start?	EN	NDP	Comparative
HST Ontario	EN	NDP	Comparative
Changer*	FR	NDP	Positive
NPD-Bloc*	FR	NDP	Negative
Travaillons	FR	NDP	Comparative
Nation*	FR	BQ	Negative
Heureusement 1*	FR	BQ	Positive
Parlons	FR	BQ	Comparative

NOTE: All ads ran in 2011 except for those ads indicated with an asterisk, which ran in 2008.

were anglophone. All were between nineteen and thirty-one years old, with a mean age of twenty-three. They reported various partisan affiliations: CPC (two), LPC (fourteen), NDP (nine), and BQ (two), with four indicating they were undecided about their vote in the upcoming federal election. This sample is rather small and non-representative because the study was conducted as part of the pre-tests of an upcoming larger study.

Experimental sessions were conducted individually at McGill University in a room arranged for this purpose. Each session began with a brief presentation of the sequence of the experiment, informing the participant that its goal was to explore short-term reactions to political ads. Participants were told they were to watch nine ads in their native language on a computer screen. The ads were randomly drawn from a sample of twelve (in each language), and shown in random order.

After viewing each ad, participants engaged in a written thought-listing task. They were asked not to censor their thoughts and not to worry about the quality of their writing. Classic studies grounded in the cognitive responses approach allow between one and three minutes for this task. Accordingly, participants were given sixty seconds after each ad to write down their thoughts. Each session ended with the presentation of a short questionnaire about age, gender, and voting intentions for the upcoming federal election.

An essential component of the cognitive response approach chosen as my theoretical and methodological framework is the concept of cognitive elaboration, central in Petty and Cacioppo's ELM (1981). In order to grasp this notion, the spontaneous cognitive responses technique (or thought-listing task) has proved to be the method of choice. It requires the receiver to verbalize – either orally or written – the thoughts prompted by a message during the exposition phase. These thoughts are then submitted to a content analysis.

As mentioned earlier, standard categorizations of thoughts do not entirely take into account the strength of cognitive elaboration. In fact, I argue that the quantity of favourable thoughts evoked by a message cannot alone be used as a measure of cognitive elaboration, especially when thoughts are coded in a binary way (i.e., +1 for favourable thoughts and –1 for unfavourable thoughts), as is the case in many studies. I regard cognitive elaboration as the result of the quantity of individuals' cognitions and, foremost, the quality of those thoughts, regardless of their valence. For this reason, and also based on the lack of distinction found in the literature between different types of cognitions, I elaborated an innovative categorization tool for the content analysis of cognitive responses.

Partially drawing on categories previously implemented by Wright, I extended the categorization system by adding categories in order to create a more detailed view of the types of responses that may be elicited, therefore allowing for a better understanding of political advertising processing. These categories vary, from a simple recall of an element of a message to a firm intention of voting for or against one candidate or party; other categories are between the two, namely judgments, corroborations (or support arguments), counterarguments, and connections (i.e., establishing a link between an aspect of the ad and one's own life).

Moreover, I subdivided the judgments by adding to them a value effect (intended or unwanted). Indeed, certain judgments, although apparently positive (e.g., "This ad makes me laugh"), run counter to the desired effect of the message, and thus labelled with a negative value of "unwanted effect." Corroborations and counterarguments were subdivided in two types: standard and elaborate, depending on the extent of the argumentation (e.g., a corroboration or counterargument supported by more than one argument or element was qualified as elaborate). Voting intentions were also subdivided according to their strength (low/moderate and high). Finally, negative values (applying to the categories of judgments, connections, and intentions) correspond to statements that are indicative of persuasive resistance. Positive values, on the other hand, relate to message acceptance.

I also took into account the strength of cognitive elaboration by attributing to the different types of cognitions a value ranging from one to ten (see Table 3.2). The value attributed is based on the magnitude of the mental effort that I inferred from the definition of each category (i.e., it seems logical to give less importance to a mere recall of an ad's element than to a firm voting intention). This value allowed me to compute a mean score of cognitive elaboration for each ad. The further the score is from zero, whether positive (favourable elaboration to persuasion) or negative (unfavourable elaboration to persuasion), the more important is the cognitive effort.

To verify part of H2, a second and much simpler categorization system was used to distinguish cognitions related to emotional/peripheral cues (e.g., music, sound effects, visual effects, colour, tone) from those related to verbal/central arguments (e.g., message's theme, type of argument). Thus, whenever a statement touched one of these categories, it was identified as one occurrence (+1), allowing me to count the frequency for each type of cue.

RESULTS

The twenty-four TV spots generated a total of 868 cognitive responses relevant to my categorization systems. For validation purposes, a sample of the statements ($n = 105$) was subjected to inter-rater reliability measures. Types of cognitive responses were given rigorous operational definitions, and two research assistants were trained in the application of these definitions before the coding. Cohen's kappa, a widely accepted statistic (Leclerc and Dassa 2010), was computed. According to the classification interpretative guidelines suggested by Landis and Koch (1977, in Leclerc and Dassa (2010), kappa values of .40 to .60 represent moderate agreement; .60 to .80, substantial agreement; and .80 to 1.00, almost perfect agreement. A kappa value of

Table 3.2

Type of cognitions, value, and example

Type of cognitions	Value	Example*
Recall	1	"red flag"; "mustache"
Positive judgment (intended)	2.5	"Canadian leadership is a good line for the NDP"
Positive judgment (unwanted)	−2.5	"Hahaha, very funny."
Negative judgment (intended)	2.5	"Taxes on iPods seem ridiculous."
Negative judgment (unwanted)	−2.5	"Fighter jets are not Canadian."
Standard corroboration	3.5	"Promoting the idea that Canadians need someone they can trust."
Elaborate corroboration	5.0	"Young-people jobs: finally someone cares! The ad really highlights young people's needs."
Standard counterargument	−3.5	"Pandering to emotional mathematically incompetent people is not good leadership."
Elaborate counterargument	−5.0	"Pretty ridiculous that Ignatieff would want to form a coalition government with the Bloc Quebecois. These two don't share the same agenda at all!"
Positive connection	4.0	"I have a family too."
Negative connection	−4.0	"This does not concern me."
Positive voting intention (low/moderate)	8.0	"I don't know who I'm gonna vote for. Maybe NDP."
Negative voting intention (low/moderate)	−8.0	"This time, I don't think of voting Liberal."
Positive voting intention (high)	10.0	"I already have decided to vote for them."
Negative voting intention (high)	−10.0	"After this, I want nothing to do with a Conservative government."

* Examples are comments generated among the participants of our study.

.72 was achieved for the overall coding. Differences in coding were resolved by discussion.

As shown in Table 3.3, the negative ads generated significantly more negative judgments than did comparative and positive ads [$F_{(2, 24)} = 7.379$, $p < .004$], whereas the positive ads generated more positive judgments than did the other two types of ads [$F_{(2, 24)} = 6.914$, $p < .005$]. Also, negative ads

TABLE 3.3

Mean cognitive responses according to advertising tone

Statement type	Condition	Mean p/ad	F value	p value
Negative judgment (intended)	Negative	4.80		
	Comparative	1.25	7.379	0.004[a]
	Positive	0.20		
Positive judgment (intended)	Negative	3.90		
	Comparative	5.50	6.914	0.005[a]
	Positive	10.30		
Counterargument	Negative	4.00		
	Comparative	1.75	2.738	0.088[b]
	Positive	1.40		

a Statistically significant at the .01 level.
b Statistically significant at the .10 level.

generated more counterarguments – both standard and elaborate – ($\mu = 4$ per/ad) than did positive ads ($\mu = 1.4$ per/ad) and comparative ads ($\mu = 1.75$ per/ad), at a 0.1 level of significance [$F (2, 24) = 2.74, p > .088$]. As for other relevant categories, findings were either not statistically significant, or the number of occurrences was too low to allow for statistical testing. Based on these findings, it follows that negative ads were generally associated with a greater resistance to persuasion, supporting H1 and an entire body of research on the negativity bias.

Interestingly, comparative ads were processed with the highest cognitive elaboration ($\mu = 2, 35$) compared with negative ads ($\mu = 1.65$) and positive ads ($\mu = 1.00$) [$F (2, 24) = 5.247, p < .006$]. This suggests that contrasts between alternatives may prompt voters to engage in more mental activity than do non-comparative ads.

As expected, cognitive effort exerted when viewing a political ad remained quite low. Most cognitive responses consisted of recalls (24.9 percent) and judgments (54.7 percent), two categories associated with low cognitive elaboration, compared with categories of moderate to high cognitive effort, such as corroborations (10.7 percent) and intentions (2.5 percent). These findings partially support H2. Regarding the type of cues prompted, I expected, based on Petty and Cacioppo's ELM (1981), that a low cognitive effort would lead to a peripheral processing of mostly superficial cues, such as audio-visual elements. A total of 136 comments were

TABLE 3.4

Mean cognitive elaboration according to party affiliation and party represented

Party represented	Mean cognitive elaboration	t value	p value
Conforms to political affiliation	3.01	2.943	0.006[a]
Does not conform to political affiliation	−0.05		

a Statistically significant at the .01 level.

made about emotional cues, particularly regarding music (46). However, this type of cue did not generate as many comments per ad ($\mu = 5.67$) as did the verbal/written arguments ($\mu = 9.17$) ($t = 8.99, p < .000$) – a type of cue associated with the central/high cognitive effort in the ELM. These findings do not support part of H2, but they could help revise Petty and Cacioppo's theoretical model by showing that it is possible to examine central arguments with a relatively low cognitive effort.

Because of a higher personal relevance of ads that match the partisan affiliation of the participants, I predicted a greater cognitive effort in relation to those ads. As Table 3.4 shows, the data does support H3. The cognitive effort to process the information provided in the ads was significantly higher and more conducive to persuasion ($t = 2.943, p < .006$) when participants were exposed to messages that reflected their political affiliation ($\mu = 3.01$) compared with when they were exposed to ads from other parties ($\mu = -.05$).

In summary, my experiment shows that the tone of political ads clearly influences the profile of thoughts generated during viewing. Negative ads generated the most negative thoughts – mainly judgments and counter-arguments – and resistance to persuasion, thus supporting work on negativity bias. An interesting result that would benefit from further investigation is the tendency of comparative ads to elicit the most mental activity. Moreover, participants engaged in a higher mental activity when exposed to ads that match their partisan affiliation, though cognitive elaboration generally remained low throughout the experiment. Finally, thoughts that relate to verbal cues were more frequently mentioned than thoughts relating to emotional cues. This suggests that a low-effort processing of central elements is possible.

Political Communication in Canada
Televised political advertising has become the dominating element of all political promotion strategies (Falkowski and Cwalina 2012). Despite the

growing body of research – mainly American – devoted to the study of its effects, there remain debates and gaps in the knowledge regarding the impact of political ads. As Goldstein and Ridout (2004, 206) point out, one of the main concerns relates to election outcomes: "For the vast subfield of voting behavior and elections, determining whether political campaigns influence individual vote choice and election outcomes has become a Holy Grail. Yet there has been relatively little work specifically on whether advertising wins elections." I argue, however, that in order to determine if and how political advertising reflects on an election outcome, it is important to grasp its short-term effects not only on voters' perceptions and attitudes about personal qualities and issue positions of the candidates but also on voters' emotions provoked by ads, and – as I did here – the way receivers cognitively elaborate about them. These impacts act as intermediary variables in the persuasion process and thus are likely to influence the ultimate voting decision. During an election campaign, voters are bombarded with TV ads, precisely because of their proven short-term effectiveness, therefore reinforcing the relevance to study their immediate effects. Since there is no consensus regarding these effects – this being mainly explained by measurement issues – more studies are needed in order to reach a larger agreement.

So far in Canada, no other research had investigated the role of cognitive responses in political advertising in an electoral context, nor in an experimental lab setting. I believe this study offers a contribution to both scientific knowledge development and political communication practice. For political advertising research, it shows that there is a distinction to be drawn between different tones of TV spots and that this difference may be important for information processing. Moreover, cognitive elaboration is influenced not only by tone – comparative ads evoking the highest mental effort – but also by political affiliation. For practitioners, the study indicates that comparative ads might be an efficient tool to communicate opponent negatives in a way that meets with more positive cognitive responses and provokes deeper thought. However, reproducing the research using a larger sample would be necessary to confirm these findings.

So-called American-style campaigning and political advertising has become widespread in Canada, as the growing prevalence of negative campaigning shows. This Americanization of Canadian campaigns (Nesbitt-Larking 2009) calls for a better understanding of its effects on Canadian voters, particularly in a context where individuals are generally reluctant to negativity, as evidenced by work on the negativity bias. However, Canadian scholars draw some distinctions with the American context; for instance, Fletcher and MacDermid (1998) note that Canadian political ads are more party-centred. Nesbitt-Larking and Rose (2004) argue that although English Canadian political culture shares a language with its American counterpart, Canadians are more resistant to the superficial slickness of political ads. Nesbitt-Larking (2009; 11) also points out that the negativity in Canadian ads is "normally contrastive and substantive, rather than being personally vindictive and mean-spirited." However, a look at the 2008 and 2011 federal election campaigns – particularly the 2011 campaign – does not support this statement, at least regarding the contrasting aspect of ads.

Indeed, the tone of political ads broadcast during the 2011 federal election campaign is undoubtedly polarized, with only 10.7 percent of all ads being contrasted. Yet, as outlined previously and as supported by Meirick's findings (2002), comparative ads prompt more positive thoughts and mental effort – conducing to a more likely message acceptance – than negative ads, which tend to evoke more negative judgments and counterarguments. This is only one of many possible short-term effects of TV spots, but in a research field where comparative ads have not yet been given much attention, these results should encourage further investigation.

Political advertising's primary role is – as it is for the media in general – to inform citizens, allowing them to better identify their interests and to participate in the democratic process. However, the great risk associated with political ads is in the quality of information presented to voters and the ability of citizens to engage in informed decision making (West 2009). With no strict regulation of the content of political advertising in Canada and with powerful advertising tools at candidates' disposal – especially when well funded – citizens are exposed to powerful and sometimes misleading campaign strategies. In a thirty-second spot and no time for contextualizing, it can be tempting to strategically mislead viewers or distort an opponent's record. It follows that political advertising, particularly when negative, does not always serve democracy, as the information is often inaccurate and misleading (Allen and Stevens 2010).

In turn, negative political ads can be beneficial for democracy. Indeed, research reveals that some negativity actually provides a catalyst for political information consumption and political participation (Goldstein and Freedman 2002; Kaid 2012). For political ads to stimulate voters' information seeking and, ultimately, voter turnout, they evidently need to present an accurate image of the candidate, opponent, or party, an undertaking that is often regarded – in a competitive context such as an election campaign – as utopian.

Political ads are not the only means by which voters get information. In fact, the weight of ads remains modest compared with other means, such as newspaper stories, debates, and television news stories, all of which are the most important information sources for voters (West 2009). Campaign ads cannot be studied in isolation from the overall context in which voters make decisions. The way media cover ads, what journalists say about the candidates, and debate performance all influence an election outcome. The receiver is now more than ever exposed to countless information of all types. In this general media context, trying to isolate televised political ads' specific impacts will remain an interesting challenge for scholars.

NOTES

1   Wright focused on three types of responses: (1) *counterarguments* are activated when the information is discrepant with the receiver's belief system; (2) *support arguments* (or *corroborations*) are in turn activated when the information is congruent with the receiver's belief system, giving way to a possible message acceptance; and (3) *source derogations* are indicative of resistance to persuasion and may be activated when the source is perceived as biased.

4

The Branding of a Prime Minister: Digital Information Subsidies and the Image Management of Stephen Harper

Alex Marland

As Chapter 2 demonstrated, political parties wage a permanent campaign of constant communications, especially if they have access to government resources. Increasingly this is designed to manage the brand image of the party and its leader vis-à-vis that of opponents. A "brand" is a non-physical communications entity; it is the sum of all media impressions in a consumer's mind that over time become a collection of evoked symbols and heuristic cues. Brands are thus some mental combination of communication processes and market positioning. Branding seeks to add value to a product or service so that a consumer develops an emotional loyalty to that choice over the alternatives. Furthermore, a branding strategy strives for communication synergies through consistent and simple messages across media platforms to reinforce impressions (de Chernatony and Riley 1998; Wood 2000).

Ken Cosgrove (2007, 7) has observed that, in politics, branding practitioners project a consistent message to select audience segments using "limited two-way communication featuring a lot of emotion, strong language, and potent pictures but little discussion of substance." As public figures, politicians develop a brand that is a combination of their actual personality, their own communication, messages from opponents, and a mediated image crafted by the news media. A leader's public persona therefore becomes what audiences decode from a barrage of mediated communication. A "party brand" encompasses the party name, logo, and colour schemes, as well as its history of policies and leaders, whereas a "personal brand" is the image of a politician (Scammell 2007; Smith and French 2009; see also Chapters 5 and 9 in this book). Branding a leader requires that public relations (PR) staff engage in reputation management by emphasizing some of that person's characteristics while concealing others. They must also protect and defend that image from potential harm while – and this is unique to politics – openly inflicting damage on their opponents' images. In theory, journalists try to see past these communication games, but in practice, they may become conduits of brand messaging. A leader's brand is thus an artificial construct that is based on a human being but which is a public entity that is superficial, manipulated, and open to interpretation.

When leaders respond to the risk of media bias by controlling their image and insulating themselves from scrutiny, we would hope that the fourth estate would counter through dogged investigative reporting. However, journalists are pressured by their employers' concern for profit and by the constant need to produce (Curtin 1999). They have grown dependent on information subsidies that fulfill the need for cheap, effortless, and timely content. For instance, press releases are conventional PR tools for presenting information, increasing the likelihood of news coverage, and influencing how subjects are framed in the media. Visuals are offered via pseudo-events, the simulated realities staged by PR personnel that provide journalists with photo ops, which are then presented to the public as spontaneous events even though they were planned and packaged (Boorstin 1992). The demand for these subsidies has increased because news organizations, especially newspapers, are in a state of financial crisis, grappling as they are with an eroding customer base as the popularity of speciality and online media increases (Sauvageau 2012). At the very time that there is a pressing need for investigative reporting to uncover *original* information about politics and government, its commercial value has plummeted, and journalists are left to compete with trends in audience ratings that signal a preference for personalities over policy, human drama over context, and visuals over text (Van Zoonen 2005; Baum 2007).

The new media environment operates in an information vacuum whose growing appetite can be filled through digital media. The e-politics era features email and text messaging; social networks such as Facebook; the sharing of visual files through websites such as Flickr, Pinterest, and YouTube; and microblogging on Twitter and Tumblr. These sources of free, accessible, and speedy information are increasingly monitored by journalists; moreover, they shift gatekeeping power away from the mainstream media and are cost-effective tools for personal branding. A growing PR information subsidy is the production and dissemination of digital information, particularly visuals. For instance, non-profit foundations and businesses seek publicity by issuing digital photographs, though this tactic is better suited for developing a relationship with journalists and for building a brand (Walters and Wang 2011). This creates an authenticity paradox of carefully constructed personal brands that are perceived as representing real identities (Guthey and Jackson 2005). Nevertheless, given that press releases that include a photo or video are considerably more likely to be viewed (Sebastian 2012), the distribution of digital visuals has quickly become a norm in the practice of media relations.

How is digital media used as an information subsidy in Canadian politics? One tactic, which is the focus of this chapter, is the circulation of controlled photographs via media releases, emails, websites, and social media. Political photographs are designed by image handlers to frame a leader in a favourable manner with the knowledge that audiences will form impressions about a politician's "demeanour, competence, leadership ability, attractiveness, likeability, and integrity" (Rosenberg et al. 2001, 123). These image bites are an excellent indicator of what political marketers want a leader's personal brand to be (Marland 2012). For instance, on June 25, 2010, the White House issued a photo of world leaders clustered around a jovial President Obama, and only the top of Prime Minister Stephen Harper's head was visible. By comparison, on the same day, the PMO released its own photo of Harper in a private meeting with Obama, who was listening intently to the Canadian leader.

The evolution of technology has led to another digital media tactic: the creation of video magazines. In 2014, as this chapter was going to print, the PMO had begun to supplement its digital photos with weekly "Stephen Harper: 24 Seven" productions. The behind-the-scenes video features a brief (under four minutes) compilation of selected content documented by PMO videographers in the style of reality television. Viewers are informed about the many events that the busy prime minister had recently participated in; furthermore, they are offered a privileged insider's view of unscripted backstage moments that would otherwise be off-limits. Each video begins and ends with "The Maple Leaf Forever" theme music. A narrator, in a journalistic tone, describes the depicted activities, such as Harper making formal announcements, meeting with world leaders, or travelling to events. Visuals of official business were interspersed with political and private moments, such as the prime minister mingling with senior citizens at a retirement home, attending a hockey game with his son, and speaking off the cuff with his staff. Each week, the attentive public are alerted via email and social media to the availability of the newest "Stephen Harper: 24 Seven" video. The tactic appears to have been modelled after the White House's "West Wing Week," which was launched by the Obama administration in mid-2010.

Such image bytes are a component of branding, one which, according to Cosgrove (2007, 14), "is a visual phenomenon ... that puts pictures into the heads of target audiences." What content do controlled political images contain? What do they say about a leader's brand? To provide baseline answers, this chapter studies digital photographs issued by the PMO in 2010 with an objective of identifying how the Stephen Harper brand was visually positioned by his staff. This provides indications of the Harper administration's policy priorities and marketing strategies, and about the prime minister's leadership. It adds to our understanding of information subsidies, image management techniques, and the "Harper brand," while spurring questions about the use of public funds for propaganda tactics.

Case Study

The prime minister's media personnel actively consider visuals in their communication and event planning. Internal instruments such as the Message Event Proposal (MEP) form prepared by departments are a mechanism for the central agency to provide direction on a media event, spokesperson, key messages, attire, and the ideal photograph. Photojournalists are sometimes bussed to an undisclosed location where a pseudo-event takes place and where they are not allowed to ask questions. The PMO employs two official photographers, who have exclusive access to backstage moments. These official photos are selectively used by political staff and, as government property, are eventually deposited in the national archives.

Although we can shed light on branding tactics, there is no definitive way to measure a personal brand. Evaluating a static brand, let alone a human brand, is such a complex proposition that marketing scholars grapple with measurement methods (e.g., Low and Lamb Jr. 2000) and theoretical constructs (e.g., de Chernatony and Riley 1998). The different measures used by international brand consultants (e.g., Interbrand 2010) are not applicable either. Nevertheless, we would anticipate certain brand-image commonalities no matter who occupies the Office of the Prime Minister. Leadership evaluation literature suggests that electors assess candidates on a wide range of criteria, such as perceived skills, personal integrity, empathy, and ideological orientation, and that the most salient characteristics vary over time (Funk 1999). By virtue of holding the position, a PM's personal brand should share characteristics with Canadians given that electors tend to prefer a leader whose visual traits are most similar to their own (Capara et al. 2007).

We would therefore expect that a successful leader's brand shares some similarities with successful domestic brands. The top companies on a list of "best" Canadian brands are foremost financial services and media services businesses whose image is generally staid, safe, and conservative (Interbrand 2010). The nationalistic brands of Molson Canadian beer (MacGregor 2003), Roots clothing (Carstairs 2006), and Tim Hortons coffee (Cormack 2012) indicate that a Canadian brand idealism portrays friendliness, collectivism, national unity, cheeky anti-Americanism, a lack of metropolitan preciousness, outdoor ruggedness, ice sports – particularly hockey and curling – peaceful law enforcement, and ambassadorship. Those businesses employ brand image transfer (BIT) whereby associations with one image are transferred to another in an attempt to change perceptions of the brand (G. Smith 2005). Their BIT marketing uses visuals of Canadian symbols – beavers, canoes, maple leaves, the flag, and toques. However, these are but some of many components of a handful of Canadian brands, and branding a political leader is not synonymous with branding a commercial product.

To varying degrees, these characteristics are present in the Stephen Harper brand. Political scientists have determined that his image as a safe economic manager has been Harper's foremost competitive advantage (Clarke, Scotto, and Kornberg 2011; Gidengil et al. 2012). In the media, he is framed as standing for "family values, law and order, pride in the armed forces, the great north, the sport of hockey, lower taxation, [and] the monarchy" (L. Martin 2010, 269). Party strategists and polling data indicate that his brand is designed to reinforce core supporters such as western populists and anglophone Protestants; to reach out to swing voters such as those in working-class families, new Canadians, and conservative racial and religious minorities; and to appeal to geographic clusters in British Columbia, Quebec, and suburban Ontario (Flanagan 2009, 223-25, 279-80; Harris Decima 2010). The Conservative Party of Canada (2011b) notes that Harper is interested in hockey, curling, music, and cats. What the Harper brand does *not* represent defines it too. In 2010, the Conservative Party's areas of vulnerability were pensions, health care, democratic reform, climate change, and helping the elderly (Kennedy 2010). It is also highly unlikely that PMO photos would involve the Conservative Party's undesired wedge issues – depicting Harper participating in a gay pride parade or addressing a labour union, for example.

METHOD

This study looks at photographs taken by the PMO's photographers in 2010 that were uploaded to the Office of the Prime Minister's website (http:// pm.gc.ca/) and that were distributed via email, RSS news feeds, and social networking websites. Specifically, it examines postings to a PMO webpage called *Photo of the Day*, which from 2010 to 2013 featured a photograph accompanied by a short description, the date, and a credit given to the PMO photographer. The webpage title implied that a new photo was issued every day; however, photos tended to be issued only on weekdays, there was a lull in late summer, and in fifteen instances photos taken on the same day were released on consecutive days. All 227 photos issued in 2010 that were archived on the *Photo of the Day* webpage were studied. The monthly total of photos ranged from a high of 28 issued in April 2010 to only 8 in July. As the content of the photos indicates, it is significant that a mean of 21.3 photos were posted in months that the House of Commons met, as opposed to 13.8 photos in months that it did not.

Analysis of visual images is challenging because their interpretation is inherently subjective. To examine how the prime minister was visually branded by the PMO, a simple quantitative content analysis of the photos was performed. The aforementioned literature about the Harper brand, combined with the researcher's awareness of Canadian politics and political marketing, informed the inductive creation of fifty-seven dichotomous coding categories of semiotics (i.e., communication signs). Itemizing the visual presence of physical things and people avoids subjective interpretations of communication, such as whether someone is smiling or not, or whether a region of Canada is depicted. Indicators of "conservative" policy issues included the economy, corporatism, law and order, the United States, and religion. "Progressive" issues were deemed to be symbolized by health care, education, daycare, same-sex rights, unemployment, and environmentalism. Special circumstances such as the Winter Olympics and natural disasters were considered. The types of people, brand iconography, location, and other details, such as clothing and props, were studied.

To prepare for coding, the visual content of each photograph was written out, as per Altheide (1996) and Penn (2000), by three political science students. Text, such as slogans within the photographs, was documented, and the names of depicted cabinet ministers were noted.[1] Each student then independently reviewed the entire corpus and added observations where necessary until agreement was achieved on a summary of approximately

TABLE 4.1

Examples of PMO *Photo of the Day* captions and coder descriptions of photos

PMO photo caption (verbatim)	Data added through coders' written description
Prime Minister Stephen Harper works from his Langevin office. March 5, 2010. Photo by Deb Ransom.	PM Harper is adjusting his glasses while talking on the phone in his office. There is paper on the table and a pen in his shirt pocket. He is wearing business attire.
Prime Minister Stephen Harper plays with foster kittens at 24 Sussex. May 1, 2010. Photo by Deb Ransom.	PM Harper is sitting on a couch with a kitten on his right shoulder and another in his left hand. There is a variety of music equipment in the background. He is wearing jeans, a collared shirt, and a suit jacket, and there is a pin of the Canadian flag on his left side.
As part of Project North's efforts to make physical fitness and sport more accessible to youth in Canada's North, Laureen Harper helps load donated hockey equipment bound for Hall Beach, Nunavut. November 1, 2010. Photo by Herman Cheung.	Using both hands, Mrs. Harper is carrying a large blue hockey bag adorned with Bauer and Nike swoosh logos. She is wearing a blue jacket and a colourful scarf. A boy behind her appears to be Aboriginal and is wearing a red jersey with an NHLPA Goals and Dreams logo. On an easel is a child's painting of an inukshuk and the words "Project North."

SOURCE OF PHOTO CAPTIONS: Reprinted from Office of the Prime Minister (2013).

twenty-nine thousand words total, inclusive of the PMO captions that accompanied each photo of the day (Table 4.1), to provide a source of information for the identification of content variables. A fourth student joined the others to separately code the presence of the fifty-seven variables in each of the 227 photographs and its written summary. On most variables, 100 percent inter-coder reliability was achieved; over 90 percent reliability occurred with other variables, which usually involved ascertaining the identity of a person or examining partially obscured subject matter. In the event of disagreement, I adjudicated.

In 2014, the *Photo of the Day* webpages continued to be archived on the PMO website, but the digital photos will presumably vanish from that site when Harper is replaced as prime minister, if not before. My formal request to the Government of Canada to reproduce select PMO photos in this chapter was denied. The reason given was that "the pictures of the day are for online viewing only" (Public Works and Government Services Canada 2013).

FINDINGS

The overarching observation from these studied photos is that the dominant brand frame that the PMO projected of Stephen Harper was of a hard-working head of government. In 82.8 percent of the photos, he was wearing business clothing, and nearly half (46.3 percent) of the photos depicted the Parliament Buildings (Table 4.2). A common visual was of Harper in the Prime Minister's Office working at his desk or talking on the telephone. Sometimes he was photographed talking with staff, attending a meeting, or in a legislative chamber. When cabinet ministers appeared (10.6 percent of the photos), they tended to be listening to the PM and held portfolios associated with Conservative Party policy priorities, such as the economy and law and order. Minister of Defence Peter MacKay and Minister of Foreign Affairs Lawrence Cannon, who were also regional ministers in Atlantic Canada and Quebec, where the Conservatives' polling was weakest, appeared five times each, followed by other members of Harper's inner cabinet at the time (Diane Finley, Jim Flaherty, Marjory LeBreton, Stockwell Day, Tony Clement). The prime minister was also framed as a hard worker when he was not in Ottawa. He toured natural disasters, attended local events, and read documents in hotel rooms. Both at home and abroad, he was pictured with international dignitaries and leaders, including Queen Elizabeth II, sometimes at major events such as APEC, the G20, the World Economic Forum, and EU-Canada summit meetings (9.7 percent of the photos).[2] The "working" frame was sustained when photos of activities taken on the same day were issued over a number of days (i.e., Harper was not necessarily working on days that the photos were released),[3] and brand image transfer occurred when he was shown interacting with construction workers (1.8 percent).

TABLE 4.2

Visual markers of the PMO's desired Harper brand (2010)

Frequency (%)	Visual
82.8	Stephen Harper in business clothing (suit jacket, necktie, etc.)
46.3	Parliament Hill (office, legislature, hallway, Peace Tower, etc.)
35.2	Canadian flag pin in Stephen Harper's lapel
22.9	Canadian flag (excluding lapel pins)
11.0	Laureen Harper

▶

◄ TABLE 4.2

Frequency (%)	Visual
10.6	Cabinet minister(s)
9.7	Stephen Harper in casual clothing (unbuttoned shirt, jeans, etc.)
	International politics (leader, diplomat, international flag, etc.)
7.5	24 Sussex Drive, the prime minister's official residence
	Law and order (courts, military, police, jail, etc.)
7.0	Winter Olympics (Olympic rings logo, athletes, events, medals, etc.)
6.6	Hockey
6.2	Canadian wilderness (forest, lake, rural landscape, the North, etc.)
5.3	Miscellaneous Canadiana (Mounties, mascots, toques, etc.)
	Stephen Harper on the telephone
4.0 to 4.8	Pop culture celebrity (professional athlete, singer, etc.)
	Solemn events (memorials, touring natural disasters, etc.)
	Sports other than hockey (curling, skating, volleyball, etc.)
	Monarchy (Queen of Canada, governor general, etc.)
3.5	Musical instruments (piano, guitar, drums, etc.)
	Stephen Harper giving a speech
2.2 to 2.6	Rachel Harper, the Harpers' eleven-year-old daughter
	Ben Harper, the Harpers' fourteen-year-old son
	Provincial politics (premier, provincial flag, etc.)
	Steven Harper signing his autograph
	United States of America (US president, flag, etc.)
1.3 to 1.8	Cats or kittens
	Economy on signage (e.g., "Economic Action Plan")
	Construction workers
	Christian religion (Christmas tree, crucifix, church, etc.)
	Education (schools, books, classrooms, etc.)
0.4 to 0.9	Aboriginal symbols (totem poles, art, traditional clothing, etc.)
	Animals other than cats (dogs, horses, etc.)
	Coffee
	Non-Christian religions (mosque, synagogue, rabbis, etc.)
0	Beer
	Corporate business (stock exchange, money, etc.)
	Daycare (preschoolers, childcare centres, etc.)
	Environmentalism (energy efficiency, electric cars, pollution, etc.)
	Health care (hospitals, medical equipment, doctors/nurses, etc.)
	Political party logo(s)
	Unemployment (homelessness, people down on their luck, etc.)

NOTE: n = 227. Each PMO photo of the day contains multiple variables.

That this dominant image operated in unison with the Conservative Party's own communications points to the resource advantage of controlling the government. A pre-election Conservative TV spot released in early 2011, "Rising to the Challenge," showed visuals of Harper wearing a shirt and tie as he worked alone at night in the Prime Minister's Office, where he was surrounded by Canadian flags. The narrator remarked:

> Our economic recovery is still fragile. That's why we can't stop working. Working to create new jobs. Working to save existing jobs. Working to help those who need it most ... Canada is stronger, prouder, and walking tall in the world. We're in safe hands with Stephen Harper. (Conservative Party of Canada 2011a)

Visually, the message was identical to many of the PMO's photos of the day. A content analysis of PMO photos, "Stephen Harper: 24 Seven" video, and party visuals would find brand message consistencies as well.

The studied photos featured Conservative policy priorities, but it was the framing of Harper's personal brand that took precedence. Given the political importance of the economic/jobs frame, it is surprising that the economy was not given a greater profile. Economic messaging had only a limited presence (1.8 percent of the photos) and corporatism was not shown. The Tories' law-and-order message was supported by images of the military and police (7.5 percent), such as of Harper inspecting the troops, touring a Canadian Forces command centre, interacting with veterans, and working with the minister of national defence. Among symbols associated with conservatism, the monarchy (4 percent) had a more pronounced place than relations with the United States (2.2 percent) or images of Christianity (1.3 percent). Education (1.3 percent) rarely appeared, and all of the remaining "progressive" policy indicators (daycare, environmentalism, health care, same-sex issues, unemployment) were not depicted. The limited inclusion of select demographic groups is somewhat, but not exclusively, indicative of Conservative priorities given that political minorities are historically under-represented in the Canadian media (Nesbitt-Larking 2001, 347-48). The proportion of studied photos with women (27.3 percent), children (10.1 percent), non-Caucasians (7.5 percent), senior citizens (6.6 percent), Aboriginals (0.4 percent), or Aboriginal symbols (0.9 percent) is contextualized by the proportion of PMO photos that depicted journalists (7 percent).

The image bites were thus foremost a tool to brand and personalize the prime minister. Viewed through this branding lens, Harper is not only a hard worker but also the head of a nuclear family who loves his wife, takes his daughter to school, and attends his son's volleyball tournament – and does so wearing casual clothes such as jeans (9.7 percent of all photos). In roughly one photo a month (7.5 percent), the setting was 24 Sussex Drive, in an apparent attempt to humanize and personalize Harper as a family man. Christmas, Easter, and Halloween were marked, including a photo of the PM greeting trick-or-treaters. Laureen Harper had a significant presence (11 percent), and she appeared more often in the photos than did cabinet ministers or international leaders. There were a number of intimate portraits with her husband, including one of them kissing, but Mrs. Harper was typically presented as a presidential first lady. She was shown speaking with world leaders and their spouses, being interviewed, hosting a diplomatic garden party, and participating in a roundtable discussion on women's rights. The practice of the Harper administration limiting public awareness of intergovernmental meetings is illustrated in the finding that the Harpers' children appeared more often in the photos than did premiers, and that occasionally Ben (2.2 percent) or Rachel (2.6 percent) was the subject. Through these photos, the soft image of the Harper family, as with that of cats and kittens (1.8 percent), was transferred to the brand of its patriarch. Political journalist Sarah Boesveld (2013) indicated in a *National Post* report that this frame was repeated in January 2013 when the prime minister's Twitter account, @pmharper, promoted photos on Flickr and video on YouTube that documented Stephen Harper's workday. The "day in the life" images began with the prime minister eating cereal with his cat Stanley – named after the Stanley Cup – nearby, proceeded to show him going to work on Parliament Hill, and concluded with him returning home to work on a laptop with Mrs. Harper looking on (ibid.).

This use of controlled images to humanize a prime minister who has been popularly depicted as animatronic is intended to counter his lack of charisma. The importance of personalities since the emergence of television has led to "the illusion of intimacy" whereby electors connect on a parasocial level with leaders (Keeter 1987). Video turns colourful personalities into larger-than-life dynamic figures, but it also accentuates the bland character traits of people who are deemed to be dull. In the studied PMO photos, a key aspect of the personalization of Stephen Harper was the framing of his personal brand as a patriotic Canadian sports fan. Sports were depicted in 15 percent of the photos. That the Winter Olympics was held in Vancouver from February 12 to 28, during which time Parliament was prorogued, led to a number of related images (7.5 percent), including of Harper with Canadian Olympic and Paralympic athletes. But Harper also appeared in sports locker rooms; at physical fitness policy announcements; at the Calgary Stampede; and attending sporting events with members of his family. In one photo, he was shown hoisting the Grey Cup among the cheering Montreal Alouettes after they won the CFL championship. Hockey was depicted in 6.6 percent of the photos. The prime minister mingled with NHL hockey players, coaches, and referees, including Gordie Howe and Don Cherry, and watched the men's Olympic gold-medal hockey game with Wayne Gretzky. This fits into the Conservative Party's other attempts to visually associate their leader with hockey and achieve brand image transfer of its attributes, such as energy, ruggedness, and the Canadian identity. Harper often attends games, where he tends to be willing to talk with reporters about sports; the media is reminded that he is writing a book about the history of hockey, that he takes his son to the rink, and that he gives hockey jerseys to international leaders; and in the 2007 budget, the finance minister's new shoes were hockey skates. As Scherer and McDermott (2011, 110) have remarked, for the Conservative Party, "hockey remains the pre-eminent signifier of a particular 'brand' of Canadianness for ... a demographic of imagined 'ordinary Canadians.'" In this way, the off-putting image of dullness is elevated into the ordinary and sameness, which many electors can relate to.

The visual frame of Stephen Harper as a hockey aficionado not only shows a commoner image but attempts to buttress his personal brand as a Canadian patriot. In over a third (35.2 percent) of the photos, Harper had a Canadian flag pinned to his lapel, and the Canadian flag itself appeared in over a fifth (22.9 percent) of the photos, often as part of the central image frame. Canadiana appeared in a variety of other ways, such as Harper clowning around with Bonhomme Carnaval, or visuals of the outdoors (6.2 percent), including a visit to the Canadian North. This is consistent with strategists' plans for the Conservative Party to align itself visually with the Canadian flag (Marland 2012, 222; Marland and Flanagan 2013).

Aside from images of the prime minister mingling with renowned athletes, there were multiple indications of the global trend toward the convergence between politics and popular culture (e.g., Van Zoonen 2005).

Music was a minor theme, with instruments such as the piano, guitar, and drums appearing (3.5 percent), typically at 24 Sussex. Pop culture celebrities made appearances (4.8 percent), including celebrity singers Diana Krall, Chad Kroeger, and Taylor Swift. Harper was himself treated as a star. In the PMO photos, he was the focus of other people's attention. A number of times, Harper is shown giving an autograph (2.2 percent) – to a journalist, a young hockey fan, and the captain of the Ottawa Senators, among others. Three photos show the prime minister playing the piano; another, him signing a guitar.

The PMO photos also document a subtle change in Stephen Harper's personal brand image: he began wearing eyeglasses in public. From the first photo of the day on January 6 through to July 9 (n = 139), he was shown wearing glasses only ten times and did so almost exclusively in private settings in his office. Then, on July 10, he was depicted laughing at the Calgary Stampede while wearing a cowboy hat, a plaid shirt, and eyeglasses. This particular PMO photo appears to have been a transition point in Harper's personal brand and a deliberate counter to a notorious photo taken by the Canadian Press at the Stampede in July 2005 in which Harper was sporting a cowboy hat, a leather vest, a string tie, and an odd expression. From that point forward, in *all* photos of the day (n = 88) in 2010, except two taken in Ukraine in October, whenever Harper appeared he was wearing spectacles. This image change likely improved public perceptions of Harper's intellectualism and integrity, if the social psychology research about people wearing eyeglasses (Hellström and Tekle 1994) is any indication.

Finally, on days that there were major political controversies (as identified by Clark et al. 2010), the PMO's information subsidies were impervious to outside events and perhaps even attempted to counteract negative media coverage. For instance, on January 23, photojournalists documented mass protests denouncing Harper's decision to prorogue Parliament (Figure 4.1). No PMO photo was issued; however, the next day it depicted Harper working with staff in his Centre Block office. On April 27, the Speaker ruled against the executive branch's withholding of documents about the torturing of Afghan detainees; the PMO promoted a photo of Harper shaking hands with an Iranian human rights activist and Nobel Peace Prize winner. On June 27, there were over four hundred arrests during violent demonstrations in Toronto concerning the G20 summit meetings, and the PMO issued a photo of the Harpers kissing. From July 20 to August 11, there was sustained criticism over the government's cancellation of the mandatory long-form

FIGURE 4.1   Example of photojournalist media photo. Photo by Chris Young,
The Canadian Press (January 23, 2010). Reprinted with permission.

census; the only photo issued, on August 5, was of Harper receiving a standing ovation from Conservative MPs. The same pattern was found on days
that the Conservative government survived a budget confidence vote (March
5); when Minister Helena Guergis resigned (April 9); when news broke
that the United Arab Emirates would no longer host a Canadian airbase
and that Canada's bid for a seat on the UN Security Council had failed
(October 13); when the federal government rejected an Australian company's takeover attempt of Potash Corporation of Saskatchewan (November 3); and when the government announced that Canadian soldiers would
begin a new training role in Afghanistan (December 9). Brand inoculation
from public controversy has also been detected in "Stephen Harper: 24 Seven"
videos, including no mention of a security breach when protesters evaded
the RCMP to interrupt Harper at an event (*Huffington Post* 2014). Although
many of the depicted events were planned in advance, the choice of which
digital images to release illustrates the subjective and partisan nature of
PMO information subsidies.

This editorial subjectivity matters if and when news organizations rely
on the PMO's information subsidies. Many events in the prime minister's
calendar are presumably worthy of a public photo that otherwise would not

attract the attention of news editors. However, news organizations must be mindful that the PMO photos that are publicly issued exhibit a partisan bias. This is most significant when media controversy is likely to produce negative coverage for the brands of the party, its leader, or the government. It suggests that image bites border on propaganda, and yet the photos did not visually convey partisanship. Party logos never appeared, and there was no prevalent Tory-blue colour scheme. Images of former Conservative prime ministers John A. Macdonald and John Diefenbaker were offset by two photos of retired Liberal prime minister Jean Chrétien, including one of Chrétien and Harper together, smiling. The only visual evidence of government resources being used for partisan reasons were two photos, one of Conservative election candidates and the other of Rachel Harper wearing official Canadian Olympic apparel, whose logo was similar to that of the Conservative Party (e.g., Akin 2009).

Political Communication in Canada

The public relations tactic of circulating digital visuals as part of a branding strategy is related to a number of political communication concepts. Foremost among these are information subsidies, permanent campaigning, and propaganda, but also aestheticization, dumbing down, packaging, and technological determinism (see Lilleker 2006). To some, the PMO's use of political photography in the digital era is a source of concern. Canadian journalists are warning that citizens are getting only a "sanitized and staged version" of political events because journalists' access is suppressed and being replaced by information subsidies (Canadian Association of Journalists 2010). Image bites are a component of what one academic calls the Conservatives' backdoor "communication by stealth" (Kozolanka 2009). Within the public service, there are reports that Treasury Board of Canada Secretariat personnel have been fretting that the inclusion of photos of Stephen Harper on government websites could weaken the public's trust and confidence in those platforms (Cheadle 2011). Yet a sense of perspective is needed. Image bites are a technological evolution of tactics that have been used by past Canadian prime ministers (Levine 1993).

At least five thematic observations can be made from this case study. First, image bites are a deceptively simple marketing communications technique. They appear to be a harmless offshoot of reality TV voyeurism but in fact are a direct marketing tactic that assists with the packaging of a personal brand. Image bites communicate simple interconnected visuals that repeat and reinforce core messages. Their regularity is a component of the permanent campaign, and they are part of the trend toward a consumer model of political communication. They are also smart politics in a multilingual society.

Second, image bites insulate a head of government from investigative journalism. Journalists have noticed that, on occasion, the PMO has released a pseudo-event photo of Harper on days that he is not present in the legislature to address controversial questions (Ibbitson 2009). Technology has given PR personnel the ability to present a favourable photo or video while simultaneously limiting the opportunities for journalists to act as communication gatekeepers. This has implications for the quality of democratic government given that spin is blatant in the midst of public controversies and that image bites verge on propaganda. Meanwhile, political parties must dedicate resources to finance this communication tactic.

Third, although many Canadians have an impression of the Harper brand, it is impossible for most people to distinguish between the public image of a leader and who that person really is. It seems likely that Stephen Harper is a hard worker, if only by virtue of his position. But a "working" frame would also apply for many heads of government whose positions involve a cacophony of officialdom and unexpected events. The public's impressions of Harper overlook that image handlers publicly convey selective aspects of their boss's authentic side and that controlled communication may convey inauthentic ones. Audiences are also subjected to messages from the prime minister's opponents, who have an agenda of promoting a negative narrative. Meanwhile, the media treats its mediated image as the authentic one, though this may vary between media outlets. PMO photos thus contribute to a socially constructed identity and the authenticity paradox. To everyone but those who know him, what is "real" about the prime minister of Canada and what is branded is unclear.

Fourth, the convergence of popular culture and politics is intensifying. The presidentialization of the parliamentary system of government and the media's treatment of political leaders as CEOs have augmented the prime minister's celebrity status. There is a presumed public fascination with Harper's personal life, with the "first lady," and even with their children. The PMO's photographs position the PM as a superstar. There is repeated visual evidence of Harper being the centre of attention, including in the company of celebrities. He gives out autographs and he plays the piano. His personal brand shares several of the characteristics of major Canadian commercial brands, whether through a shared strategy of the brand image transfer of patriotic symbols such as the maple leaf and hockey, or simply by projecting an image of a dull but competent economic manager. Visuals of an economist with kittens, which risk generating ridicule and laughter, are used sporadically and intermingled with more subtle attempts at brand image transfer, such as a drum set as a background prop. As with American presidents (Mayer 2004), the executive office communicates messages that the head of government is in tune with average Canadians, even though by virtue of the position he travels in elite circles. The treatment of any political leader as a celebrity presents a problem for good government, as it contributes to deference to elites, the dumbing down of political debate, and the evaluation of leadership on the basis of personality, image frames, and propaganda.

Fifth, little is known about the media's uptake of these information subsidies, or the implications for voter behaviour. PMO photos respond to and cultivate the media's interest in the personalization of politics (see Chapter 9 in this book). Earlier research has indicated that smaller media operations, including online media, are more susceptible to reproducing PMO photos and that well-financed media organizations do so rarely (Marland 2012). Anecdotal evidence indicates that image bites are used in a variety of ways. Sometimes the photos are reproduced later for purposes other than the sponsor intended. One emerging trend is that they are a resource for journalists when political staffers become the subject of investigation. For instance, when CBC.ca reproduced a July 2012 wire story about a senior PMO aide taking a job with Air Canada, it included a PMO photo of Harper talking with the aide that had been issued on March 9, 2011 (CBC News 2012a), whereas other news websites used stock photos of Air Canada airplanes. When there was controversy in 2013 over revelations that Harper's chief of staff, Nigel Wright, had given over $90,000 to Conservative senator Mike Duffy to defray improper housing allowance claims, the media regularly used a PMO photo issued on March 18, 2011. That photo depicted Harper and Wright in a pensive mood, staring at the floor in the Prime Minister's Office. According to the PMO byline, the men had just released a statement about civil war in Libya, but this context was rarely mentioned when the photo was reproduced during the subsequent Senate scandal. Given that online content is in a constant state of flux, it is noteworthy that these photos remained on the prime minister's official website despite the unwanted attention, which adds to their salience as a source of data. Image bites are also used for background research: the *Toronto Star* confirmed that

Harper had begun wearing eyeglasses in public by consulting photos on the PMO website (J. Smith 2010). If the remarks of the *Globe and Mail*'s Ottawa bureau chief, who wrote that "Mr. Harper is the hardest-working prime minister in living memory," are any indication, image bites do succeed in influencing the prime minister's brand image (Ibbitson 2012). However, further research is needed to determine how the media uses these information subsidies. Moreover, whether this has any effect on voter choice is unclear. Leadership has less of an impact on voter behaviour than is commonly believed, and many Canadians cannot even name the prime minister or opposition leaders (Gidengil et al. 2006). It may depend not only on what the media reports but on how much exposure voters have to political news.

This still leaves questions about the Stephen Harper brand. PMO photos suggest that in 2010 he wanted his brand to be a hard-working and patriotic leader, who outside politics is a family man and, like many commoners, a fan of Canadian sports. There are also indications that his advisers wanted to project him as a revered celebrity. Defining the Harper brand is a temporal question. Readers will have to review a wider body of evidence, add their own impressions, and arrive at their own (subjective) answer. One observation is that some Canadian prime ministers become associated with visual icons that embody their public persona, such as Pearson's bowtie, Trudeau's rose, or Mulroney's Gucci shoes. If Harper's handlers are successful, one of the enduring iconographic associations with his brand will be hockey. Regardless of public impressions of Harper's personal brand, or who succeeds him as prime minister, it appears likely that as branding becomes more pervasive it will encourage the further convergence of party, personal, and government communications.

ACKNOWLEDGMENTS
The author would like to thank Memorial University research assistants Cody Cooke, Sean Fleming, Megan Sheppard, and Stephanie Roy. Earlier versions of this chapter were presented at the Academy of Marketing Conference, University of Liverpool Management School, July 5-7, 2011, and at the Canadian Association for Programs in Public Administration (CAPPA) Research Conference in Public Management and Public Policy, Ottawa, May 28-29, 2012.

NOTES
1   Text in stage backdrops, slogans, taglines, or signage appeared in 14 percent of the photos, none of which had French-only text. "Canada" was the most common word, followed by words associated with hockey or international politics.

2   Occasionally, world leaders were mentioned in the accompanying photo caption, for instance, identifying the person Harper was talking with on the telephone as the prime minister of Israel (December 3, 2010).

3   Four photos taken on April 14, 2010 – Harper playing the piano at 24 Sussex, meeting with New Zealand's prime minister, working in his office, and in the Centre Block of the Parliament Buildings – were issued as separate photos of the day.

5
Selling Social Democracy: Branding the Political Left in Canada

Jared J. Wesley and Mike Moyes

The importance of branding in the business world is beyond dispute, yet its function in other areas, including politics, is only beginning to be understood. As the focus of business literature shifts away from marketing as "selling a product" and toward "brand development," political scientists are challenged to keep pace. Branding is a long-term process that differentiates the product from its competitors and, in many instances, has become the primary focus for all business decisions (Ries and Ries 2002). Branding is about showcasing a story, lifestyle, or emotion, as Naomi Klein (2000, 51) describes in the context of the shoe industry: "Nike is the definitive story of the transcendent nineties superbrand ... This is a shoe company that is determined to unseat pro sports, the Olympics and even star athletes, to become the very definition of sports itself." In a similar fashion, political parties are challenged to develop and maintain their brand, as citizens (consumers) are seen to evaluate parties (like businesses) on this basis. Chapter 4's examination of the prime minister's brand image invites questions about how political parties use communication to define their own brands. A party's brand is its intangible image and reputation in the electoral marketplace (Scammell 2007). As in the corporate world, party brands simplify the political landscape, differentiate between otherwise similar competitors, and streamline the amount of information required for voters to make informed democratic choices. Through standardization of the party's image, successful brands forge both external trust (with voters and supporters) and internal unity (among party members and activists). At the same time, by requiring both external credibility and internal adherence, brands help link a party's electoral reputation to its value-based authenticity (Needham 2005).

Even when popular, managing a party's brand can be a challenging task, requiring flexibility and adaptability in the face of a constantly changing political environment. In cases of unpopularity, changing a party's brand is not a short-term project given the reputation it has built (or been branded with) over time. Just as companies have enduring corporate brands, so too do parties have lasting images in the minds of voters (Kavanagh 1995; White and de Chernatony 2002; Scammell 2007).

As Smith and French (2009) argue, a party brand has three key elements: its values, its leader, and its policies – the importance of which will vary from party to party, and from time to time. When one element of a party's brand becomes discredited or falls out of favour in the marketplace, successful parties will attempt to revamp that component or elevate other elements of the brand to greater prominence. Political communication is therefore key to effective brand management, as its tactics involve identifying, highlighting, framing, and presenting the most profitable elements of a party's image.

Some studies suggest that left-wing parties, like the NDP, are too wedded to principle to become brand-centred or market-oriented (Gibson and Römmele 2001). In practice and in theory, branding and political marketing are grounded in an economic view of political competition, in that they assume elections constitute "markets" and parties act as profit-maximizers in seeking to gain enough votes to secure office. By extension, market-oriented parties are those that use market intelligence to discern voter preferences throughout the campaign process, from the development of the party's brand to its sale to the electorate (Lees-Marshment 2001). Taken to its extreme, this market-oriented approach posits that parties should not only present themselves in a way that appeals to the electorate but also actually design their policies and messages according to what voters want, even if this means straying somewhat from the wishes of the party's membership, their ideological predispositions, and their previously established policies. This sits uneasily with many in the Canadian left, and those in the New Democratic Party.

This tension between electoral and ideological goals is not unique to left-wing parties, of course. Political organizations of all kinds struggle to balance the pursuit of power and influence with the promotion of principle (see, for example, Chapter 11 in this book). The challenge is complicated for left-wing parties, however, in that the principles held by certain party members are often antithetical to the market-based approach to the economy, politics, and campaigning. The notion that the electorate is akin to a marketplace, and citizens are reducible to consumers, is as unacceptable to some social democrats as is the idea that market forces should help determine social outcomes (Pettit 2009). Conservative parties face fewer constraints in this regard, as their members are typically less resistant to business techniques.

Traditionally, New Democrats have been described as being "fierce partisans with a definite point of view" (Rae 1996, 61). Their commitment to specific values may be traced to the party's roots in the early-twentieth-century farmer and labour movements, which coalesced to form the NDP's predecessor, the Co-operative Commonwealth Federation (CCF) in 1931 (Wiseman 1985). The principles of the CCF, as outlined in the Regina Manifesto in 1933, centred on the elimination of the capitalist system in order to achieve a more equal society. Wiseman (1985, 75) describes the political communication of the CCF as reflecting a "depression psychology" that used doom and gloom even in good times, in an attempt to play to the party's perceived advantages in providing economic security. This strategy, of stressing crisis and prioritizing the interests of Canada's most disadvantaged, persisted after the CCF joined forces with the Canadian Labour Congress to form the NDP in 1961 (McLaughlin 1992, 222). Even the softened language of the party's 1956 Winnipeg Declaration employed rhetoric that would be considered extreme by today's social democratic standards.

It was not until the early twenty-first century that the party began distancing itself from this conventional social democratic approach to campaign communications. As Michael Balagus, former chief of staff to the premier and Manitoba NDP campaign director, describes the shift, "Traditional NDP politics relies a lot on confrontation. It relies a lot on class. And when you do that, the tent gets smaller and smaller. I think by being more inclusive, by working with the business community ... By working with labour, by working and also broadening the tent, [you have] ... a very successful formula" (Balagus 2011). This type of thinking carried the day on April 14, 2013, when delegates to the NDP convention voted 960 to 188 to remove all references to socialism from the party's constitution. Borrowing language directly from late NDP leader Jack Layton's letter to posterity, the party now reserves a role for government in "helping to create the conditions for sustainable prosperity" by uniting "the best of the insights and objectives of Canadians who, within the social democratic and democratic socialist traditions, have worked through farmer, labour, cooperative, feminist," and other movements.

Discussed in the following pages, this "formula" aligns well with the one employed by Tony Blair's New Labour (Lees-Marshment 2001; Savigny 2005; Wring 2005; Lloyd 2009). The full story of "New Labour" is well known to most students in the field of political marketing and does not bear repeating here. Of note for present purposes, through a combination of inoculation, moderation, and simplification, Blair's three-pronged rebranding model would offer lessons for left-leaning parties throughout the Western world, including Canadian New Democrats.

Successfully rebranding a political party requires, among other things, reducing the risk perceived by voters should the party be elected (Lloyd 2006, 61-62). Known as inoculation, this process involves disassociating the party from the least popular elements of its image by directly confronting its (perceived) past transgressions. Depending on one's perspective, inoculation may appear as an exorcism, or an abandonment of a party's core beliefs. Either way, the strategy attempts to place a fresh face on the party and shift the focus away from the most negative aspects of its image. In Blair's case, inoculation involved a careful combination of policy shifts, creative rhetoric, and a deft, if simple, change of the party's name (to "*New* Labour"). Another key component of inoculation within the party was Blair's removal of clause IV from Labour's constitution. The article committed the party to establish "common ownership of the means of production, distribution and exchange" and was reflective of the period in which it was written, around the Russian Revolution in 1917 (Heath, Jowell, and Curtice 2001, 106).

Second, rebranding often involves moderation – "taking the edge off" the more extreme components of the party's image by updating the party's policies and ideals to mesh with those of mainstream society. By shifting focus away from full employment and government largesse as its ultimate goal, and comprehensive state planning as a means to that end, the emergence of the "third way" movement marked a turning-point in the history of social democracy (Kitschelt 1994, xiii; Green-Pedersen and Van Kersbergen 2002; Wesley 2011b).

Finally, rebranding in the modern age of sound-bite politics often involves a simplification of a party's platform. Gone are the days of telephone-book-sized programs, designed more for the reference and approval of the party's membership. Today's campaigns are designed to be media- and citizen-friendly, breaking down party platforms into shorter, fewer, more easily digestible "planks." Under Blair, New Labour produced succinct

FIGURE 5.1

New Democratic Party election performance, 1990-2011

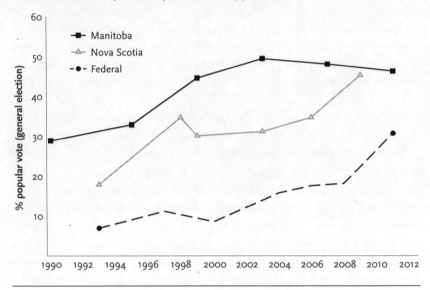

ten-point plans in place of its old hundred-page platforms (the most infamous of which was 1983's *New Hope for Britain*, which one Labour MP
once referred to as "the longest suicide note in history.") In all three prongs,
Blair was motivated by the conviction that Labour had to "adapt to the
voters, rather than persuading voters to adapt to the party" (Heath, Jowell,
and Curtice 2001, 6). In the process, New Labour helped redefine left-wing
politics in Britain and beyond.

This study reveals the influence of Blair's model on the recent success
of three New Democratic parties in Canada: "Today's NDP" in Manitoba
(whose adoption of these techniques in 1999 helped establish Canada's
longest-serving New Democratic government); Darrell Dexter's NDP in
Nova Scotia (whose adaptation of the "Doer model" helped produce Atlantic Canada's first-ever New Democratic government in 2009); and Jack
Layton's New Democrats (whose conversion to these new means of political
communication helped, in part, vault the NDP to its first term as Canada's
official opposition in 2011). For context, see the electoral results presented
in Figure 5.1.

Case Study

METHOD

The following case studies are grounded in interviews with eleven NDP advisers and strategists, including party leaders, MLAs, chiefs of staff, communications experts, pollsters, and volunteers. The interviews were conducted between April 2010 and March 2012. Their collective insight reveals the intricate relationships between ideology and policy, strategy and tactics, internal and external marketing, mobilization and persuasion, and branding and governance.

Respondents' comments were analyzed using an established three-stage qualitative coding method (see Wesley 2011a). In phase one of "open coding," researchers read through the transcripts in search of general patterns. Stage two of "axial coding" involved mining the transcripts for specific instances of these identified themes. Stage three ("selective coding") entailed reviewing the results for both intra- and inter-coder reliability. To ensure trustworthiness, findings have been triangulated using secondary literature, media reports, and election artifacts; detailed quotations have been provided to allow for verification; established interview and qualitative analytical techniques have been employed; discrepant evidence has been reported; independent interviews of a subsample of the same respondents have been conducted; member checks have been performed; and the research has been subjected to peer review. The study was approved by the University of Manitoba Research Ethics Board.

FINDINGS

Like Labour, the Manitoba New Democrats' long period in opposition helped set the stage for the party's rebranding under Gary Doer. The Manitoba NDP's rock-bottom moment came in the 1988 election, the worst performance in the party's modern history (losing nearly one-fifth of its popular vote and two-thirds of its seats). Previous analyses attribute the NDP's fall from power, in part, to its abandonment of middle-of-the-road policies (Wesley 2011a, 222-23; 2011b, 160-63). This led, as it did in Britain, to charges that the New Democrats were an old-line, tax-and-spend party. Selected on the eve of the 1988 election, NDP leader Gary Doer spent the next eleven years in opposition, attempting to reclaim this space and return his party to power.

Following a minor recovery in 1990, the Manitoba NDP began rebranding in earnest after experiencing a third consecutive defeat in the 1995 provincial election (Vogt 2010). This defeat served as a catalyst for the Manitoba NDP and made clear that "new approaches and new ways of presenting the party" were necessary (ibid.; Turnbull 2010). Fortunately for Doer and his strategists, they had the Blair playbook on which to draw.

As with Blair's repackaging of New Labour, Doer's rebranding of his party as "Today's NDP" was a key component of his inoculation strategy. According to former communications director Riva Harrison, the new name was an attempt to simplify the party's transformation: "You maybe have some impressions of us on past things, but we're asking you to let those go as we're moving forward with some new ideas. We still stand for the same principles, but we're a more modern party. Doing things differently. We're not your dad's NDP" (Harrison 2011). As a brand, Today's NDP was used exclusively by the party in all press releases and interviews during the 1999 election campaign. Likewise, the party logo was updated to include Today's NDP and was designed so that the media would not be able to separate the two words.

Like Labour, the NDP was also saddled with decades of ideological and policy baggage associated with its image as an old-line social democratic party. Between 1995 and 1999, Doer and his strategists began selling the party's membership on a bold course-change – one that would see the party adopt two new policy positions as a means of inoculation. First, the NDP would reverse its aversion to balanced budget legislation, pledging to abide by the PCs' stringent anti-deficit law (Wesley and Simpson 2011). Said NDP adviser Shauna Martin, "A lot of the balanced budget stuff and the way we talked about it, the language we used, the way we rolled out our campaigns, not just in the election but in the lead up to the election, was very much focused on how New Labour had achieved success" (S. Martin 2011). Second, Doer removed the renationalization of Manitoba Telephone System (MTS) from the NDP platform. The Progressive Conservative government had privatized MTS in 1996, a controversial move that New Democrats vehemently opposed as a sellout of one of the province's prized Crown corporations. Together with its adoption of balanced budgets, this pledge to support private enterprise marked an about-face for the NDP. Third, the NDP promised to introduce campaign finance reforms that would prohibit corporate and union donations to political parties. At the time, the proposal was seen as a slight to the role of unions in the life of the

New Democratic Party (Wesley and Stewart 2006). Whether real or perceived, this marginalization of labour was designed to distance the NDP from its image as a union-beholden party.

As in Britain, inoculation also required the NDP to moderate its approach to public policy, abandoning old-left positions in favour of pragmatic third-way principles that Doer felt more accurately reflected the ideals of Manitoba society (in Wesley 2011a, 232). Declaring that the days of "building the New Jerusalem" had ended in the 1970s, Michael Balagus noted:

> I'm not convinced that it's a more centralist position as opposed to a more effective position. And not just in terms of getting elected, but in terms of getting things done. I think the number one thing in government is you have to get things done. That's what you're here to do ... And bringing in a bunch of legislation that gets wiped out the day after you leave government and nobody misses it, is not a particularly effective way to get things done. (Balagus 2011)

This "pragmatic, not dogmatic" approach encapsulates the Doer approach to moderation.

Simplification and moderation flowed from inoculation, in that the NDP wanted to counter its critics' allegations that the party had unrealistic and extreme designs to change Manitoba society. Applying former Manitoba NDP premier Ed Schreyer's old adage that "pamphlets, not novels," win campaigns, Doer explained:

> In the past we used to have huge NDP policy weekends and would produce these fat books dedicated to policy. But this time we took a simpler approach, fearing that if we produced another 600-page document then we would almost certainly lose the election. In its place we produced five pledges and we made sure that each one, and this is perhaps a novel idea, could actually be implemented once we became the government. (Doer 2000, 5)

This shift is notable when examining the structure of Manitoba NDP platforms from 1990 to 2007 (Figure 5.2). Over time, Doer's team dramatically reduced the length of their platforms and focused on a limited number of core commitments that served as guideposts for the fulfillment of the party's mandate in office, and a sort of contract for voters.

FIGURE 5.2

Manitoba NDP platform contents, 1990-2007

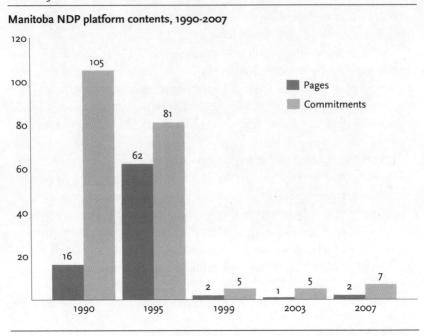

Referring explicitly to the lessons he drew from Blair's campaigns, Now Communications founder Ron Johnson noted that New Labour "was very much focused on recasting their image and narrowing down the number of things they were trying to communicate. And that's what we're doing. It's not that we're throwing policy out or abandoning things as much as narrowing what you communicate so people actually hear the things you want [them] to hear frequently enough for it to have an impact" (R. Johnson 2011).

The importance of Doer's leadership and his personification of the Today's NDP brand cannot be underestimated (Harrison 2011). The party was branded around the image of Doer and around the question of who the public thought would make the best premier. As a result, his name and face dominated most campaign materials, including all candidate lawn signs, which were emblazoned with "Gary Doer and Today's NDP." The NDP's success has transcended the departure of Doer, but with the election victory under new leader Greg Selinger, the party brand was deeply entwined with Doer's pragmatic style. Although the NDP was successful following Doer's exit, the party's 2011 campaign placed far less emphasis on

Selinger as the leader, and instead focused on local MLAs and candidates, targeted messaging on key issues (ensuring Manitoba Hydro remained publicly owned and protecting public health care), and a noticeably negative approach to attacking the official opposition Progressive Conservatives.

An electoral setback also set the stage for the Nova Scotia NDP's rebranding. For them, the 1999 election was more of a disappointment than a rock-bottom moment. A year earlier, the party had doubled its popular vote to 35 percent – the most ever earned by an Atlantic Canadian New Democratic Party. In the process, the NDP captured nineteen of the legislature's fifty-two seats – the same number as the governing Liberals. Although the Liberals were able to maintain power with the support of the Progressive Conservatives, the NDP had for the first time become the official opposition and appeared well positioned to challenge for power in the next election. The results of the 1999 election proved otherwise, however, as the Progressive Conservatives vaulted into government after the collapse of the NDP and Liberal vote.

The party turned to a new leader in 2001, and Darrell Dexter brought with him a new approach to political marketing. Dexter's model was heavily influenced by the success of the Manitoba NDP just two years earlier (S. Martin 2011). According to former Nova Scotia NDP provincial secretary Mike MacSween (2011):

> Manitoba and Nova Scotia have a lot in common in terms of population [and,] to some extent, socio-economics. We're both sort of middle of the pack, not necessarily rich provinces, but we have a lot in common. Common traditions. So we picked up on the Manitoba model as certainly being one which would work here in this province quite well. And fortunately we were correct about that.

While ultimately successful, the model used by the Nova Scotia NDP did not result in immediate success. Rather, the NDP needed to consistently apply the model over three election cycles before the party was elected to form government.

Without a previous government record, the Nova Scotia NDP did not have to "run away from" an unpopular record. To assess their own shortcomings, the Dexter New Democrats used market research to determine Nova Scotians' perceptions of their party and platform (MacSween 2011). The plan was to ensure "we didn't seem so scary," noted Shauna Martin

(2011), to "talk to [voters] meaningfully about what is perceived as a weakness, which is generally still financial issues." This explains the party's costing all of its pledges, and its commitment to "living within our means" by balancing budgets. The overall aim was to prevent the NDP from letting itself "get boxed in as a party that always wants to take money from your pocketbook" and instead inoculating itself by committing to balance the budget and reduce the debt (O'Connor 2011).

The second major component of the Dexter model involved taking a more moderate, "pragmatic approach to government" (S. Martin 2011). "Obviously we, as a party, have our firm base in the ideology of social democracy," Mike MacSween reported. "But at the same time, I think Darrell as leader and as premier was more interested in talking about solutions than in talking about ways of thinking. It goes back to the whole notion of talking about solutions, commitments, rather than talking about platitudes" (MacSween 2011). To accomplish this, the party focused its platform on the notion of "doability."

This policy shift marked a noticeable departure in both content and style, compared with the Nova Scotia NDP's 1998 and 1999 campaigns. This is particularly evident in economic issues, job creation, and the party's disposition toward corporations. In 1998, the NDP stated that the "Liberals and Tories do the bidding of the banks and big corporations," and "when the chips are down[,] they give in to the big corporations ... and the big banks" (Nova Scotia NDP 1998). This position is contrasted with a commitment in 2009 to "rewarding investment in Nova Scotia companies" (Nova Scotia NDP 2009) and completely abandoning any criticism of corporations. The NDP also reversed its position on corporate taxes; where in 1998 the party advocated increases, in 2006 it pledged to phase out the Large Corporations Tax (Nova Scotia NDP 2006). And whereas the NDP's 1998 job creation program was exclusively centred on increasing the size of the public sector, its 2009 plan involved tax incentives that would encourage the private sector to increase hiring. By making these changes, and through commitments to reducing costs for "today's families," the NDP was able to inoculate and brand itself as a moderate party.

Lastly, simplification was at the heart of the Dexter model. According to MacSween (2011),

> Rather than talking about grandiose notions and ideas and ideologies, what we talked about were very specific defined things that we,

FIGURE 5.3

Nova Scotia NDP platform contents, 1998-2009

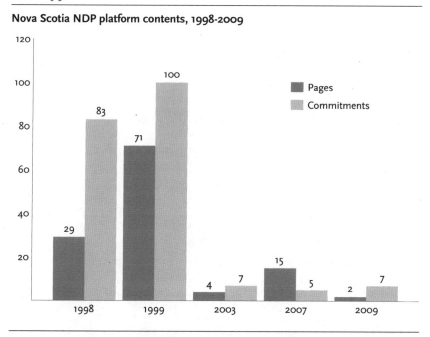

as a party were seeking to do ... The Dexter model is having a well-defined plan rather than simply "vote for us because we think differently than the other parties do."

With Robert Chisholm as leader for both the 1998 and 1999 elections, the Nova Scotia NDP campaigned on traditional left-wing platforms, containing a total of 83 and 159 commitments respectively. In 2009, Dexter's simplification resulted in a two-page leaflet: one side with a list of key platform planks, and the other with a detailed timeline and costing for those commitments. At the bottom of the page, the party invited the reader to retain the leaflet to keep track of the government's progress in fulfilling its pledges. This "contract" approach mirrored that employed by New Labour. The move to the pledge card helped simplify the party's message and demonstrated a clear break from the party's past (O'Connor 2011) (see Figure 5.3). The break from the past and the rebranding of the party fit well with Dexter's leadership style. His straight-talking approach, combined with his atypical NDP background as a former navy officer, journalist, and lawyer, played an important role in demonstrating that the Nova Scotia NDP was in fact a credible and moderate party that could be trusted to govern.

Borrowed from Blair and Doer, Dexter's three-pronged strategy appeared to pay off, as the party continually built support through 2003 and 2006, before becoming Nova Scotia's first NDP government in 2009.

The federal NDP's branding efforts were even more extensive. Before Jack Layton's leadership, the federal NDP was best classified as a product- or sales-oriented party. Following the 2000 election, "the party ... lacked resources. It lacked ... a strategic direction and it also lacked ... a modern infrastructure which included talented individuals using state-of-the-art tools to wage modern campaigns" (Lavigne 2012). The aim therefore was to "professionalize and broaden the base of the party, with a goal to put the party in a position of governing. That is, ask the country for the responsibility of governing after we had achieved a number of those interim steps that we recognized need to be taken" (ibid.). Those steps included building the financial capacity of the party and recovering from "the challenging decade of the 1990s" in order to win back the social democratic voting base and, eventually, expand on it (ibid.). According to interviews with Layton campaign advisers, the 2011 federal election was the culmination of the NDP's work over the 2004, 2006, and 2008 elections in building capacity in political marketing. This work began with Jack Layton's selection as NDP leader in 2003 and continued after his death.

Contrary to conventional wisdom, which suggests provincial parties take their cues from their federal cousins, the Layton team drew on the expertise and experiences of the Doer and Dexter campaigns (Kostyra 2011). "We were always very conscious of learning lessons from our successful provincial parties," recalls Jack Layton's chief of staff Anne McGrath (McGrath 2012). The federal election planning group included people from Manitoba and Nova Scotia. "We had to translate that into a federal context because Canada is obviously much more than just the provinces and territories that make it up. So you can't copy and paste per se, but you can certainly learn from it" (Lavigne 2012).

It was not until the NDP surge midway through the 2011 election campaign that the media (and likely many Canadians) first tuned into the New Democratic Party platform. This said, the content of the federal NDP's commitments was familiar, in that they represented an attempt to inoculate the party from its opponents' attacks. The title of the campaign platform was "Practical First Steps." This messaging included "modest, doable proposals" instead of what some would consider the pejorative connotation of the NDP starting "big, big programs which were worth billions and billions of dollars and would take years to implement" (Lavigne 2012). These short planks placed the focus on personal-level issues rather than on abstract ideology. The five key commitments were also strategically chosen in order to do one of three things: play to the NDP's strength (i.e., health care, pensions, trust), address a traditional weakness (making life more affordable), and differentiating the party on the number one issue (jobs and the economy).

Moderation was also central to the federal NDP's success in 2011. McGrane (2011) argues that the party gradually shifted to the political centre, adopting third-way-style policies, market-copying techniques in English Canada, and a market-challenging strategy in Quebec as a means of generating increased support. Like Doer, Layton downplayed the ideological nature of his campaign. Our interviews support this point; as campaign director Brad Lavigne noted, "We were looking for a niche within the marketplace. And it's not on a left-right spectrum. It's not on a federalist or sovereignist spectrum." Instead, the party emphasized Jack Layton as the most trustworthy leader to help "get things done" in Parliament (Lavigne 2012). In this vein, the party's primary means of inoculating itself against attacks based on its traditional image consisted of labelling itself a "Jack Layton party. This was done for a number of reasons," recalled Brad Lavigne.

> We knew that the leader was liked. We knew that the leader was a good contrast to the other leaders at the time. And we also knew that by branding the leader as the party, we'd be able to shed a lot of the negatives that came with the party brand. The party brand ... had plenty of negatives ... that we recognized that we could actually overcome ... by redirecting people's attention away from the party to the leader. (ibid.)

The federal NDP also implemented a familiar communication strategy. Referring to the Manitoba team, Anne McGrath (2012) noted:

> They were very focused in their messaging. They kept things fairly straightforward and they didn't put out, for instance, 200-page platforms ... They had very, very straightforward messaging that was about people and not about the party which, I think is very, very important. And they were very disciplined.

Figure 5.4

Federal NDP platform contents, 2000-11

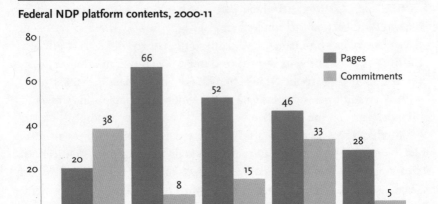

This focused and disciplined messaging can be seen in the evolution of the federal NDP platforms. Up until the 2011 election, the federal NDP campaigned on a comprehensive program. Figure 5.4 details the page length and number of commitments included in the NDP platforms from 2000 to 2011.

In 2011, the federal NDP implemented a hybrid approach to its campaign platform. Although still releasing a relatively comprehensive program, it used public messaging to focus on five key commitments that would be implemented within the first hundred days of forming government.

Political Communication in Canada

This analysis reveals several lessons for students and practitioners of left-wing campaigning in Canada. First, and it bears repeating: Doer, Dexter, and Layton built support for their campaign strategy incrementally, over several election cycles. Campaign after campaign, each leader was able to demonstrate steady, significant gains in both popular vote and seats, to the point that they were on the verge of forming government. The closer they came to power, the more each leader was able to draw on arguments that bold new approaches would be necessary to take the party's "next big step." Tony Blair's famous maxim – "Power without principle is barren, but principle without power is futile" – became an internal marketing slogan for each of the three Canadian NDP leaders as they sold their plans to the membership. Without the credibility of having raised their parties' fortunes to new heights, and without the prospect of being able to deliver on their promises of power, however, it is unlikely that Doer, Dexter, or Layton would have derived the internal support necessary to employ their respective models (MacSween 2011). Said market researcher Leslie Turnbull,

> When you go to win, you basically say to the party "look, you just have to have confidence in us. We are going to win. We can't do what you want to do in terms of ensuring regular increases in the minimum wage, reducing disparities, trying to bring up gradua- tion rates amongst young people in the inner city ... If we don't win, none of the stuff you want matters. (Turnbull 2011)

Second, although more difficult to measure, each party also benefited from a popular leader. Doer, Dexter, and Layton were all known for their Everyman character. This appeal was important not simply in terms of sell- ing the party to the broader electorate but also in terms of selling the party on a new campaign strategy. In this sense, the long tenure of these three leaders also helped them build the trust necessary to convince members to move forward with their chosen course of action. In his analysis of the Layton New Democrats, McGrane (2011) found it surprising that "there was almost no public criticism of the party from unions, local party members, or candidates" in reaction to the party's new campaign course, as all appeared content with its pursuit of power. Earlier studies of the Doer New Demo- crats revealed similar internal reaction to the party's ascendancy (Wesley 2006). In this sense, each of the three leaders was able to rebrand his party to appeal to his electorate's mainstream voters, while at the same time maintain the support of his party's central clientele, including labour unions. Unlike New Labour, the Manitoba, Nova Scotia, and federal NDP did not com- pletely sever the formal and informal ties with labour unions as a means of inoculating themselves. Rather, the changes to the NDP have been described as an "evolution" that allowed any dissension or discomfort toward the branding exercises to be controlled internally (R. Johnson 2011).

Third, each party benefited from its informal connections with counter- parts elsewhere in Canada. Compared with other Canadian parties, the New Democrats have a relatively cooperative and reciprocal relationship among their various federal and provincial wings. Officially, the federal NDP is a federation of provincial parties, which includes having a shared member- ship (Wiseman 1985, 74). Although there are few formal mechanisms in place to coordinate or organize the various provincial organizations, strong peer relationships persist and thrive among many of them. Certain institutions help facilitate these connections, including the federal NDP council, with widespread representation from the regions and provinces; the annual federal convention, whose delegates often include key party organizers from the different provinces and federal party; the Election Planning Committee that is formed prior to each election and includes NDP members from across the country; and the widespread use of the same political marketing (Now Communications) and market research firms (Viewpoints Research). These linkages help facilitate the spread of effective political communication techniques across the country.

Fourth, there are obvious parallels between the strategies and tactics employed by the New Democrats and those of their federal Conservative opponents. Discussed elsewhere in this book, the most obvious similarity lies in the two parties' use of moderate, simplified messaging. This strategy involves the use of pragmatic rhetoric and a limited amount of commitments in order to enhance voter recognition, comprehension, and recall. The 2006, 2008, and 2011 Conservative platforms, like the various recent NDP platforms described in this study, placed a large emphasis on five to six easily digestible and deliverable promises. Each list contained pledges designed to protect the party against its greatest weaknesses. For the Conservatives, these included the allegation that Stephen Harper had a hidden agenda, and for the NDP, that the party was not moderate, practical, or fiscally responsible enough to govern. As discussed in this chapter, New Democrats refer to this approach as "inoculation." Conservatives, on the other hand, call it "triangulation," – "the acceptance of deeply rooted popular views, even when they were originally associated with opposing parties" (Flanagan 2007, 78). These parallels are not coincidental. Numerous federal and Nova Scotia strategists reported learning from the Harper Conservatives' successes and adapting these strategies and tactics – along with Doer's and Blair's – to suit their own campaigns.

In sum, the resurgence of the Manitoba NDP in 1999, the Nova Scotia New Democrats' breakthrough in 2009, and the ascendancy of the federal NDP in 2011 all have several elements in common. All three campaigns involved inoculating the party against charges that it was bent on tax-and-spend policies; demonstrating the party's conversion to a new, more moderate approach to social democracy; and a concerted simplification of the party's platform.

Their success provides a series of poignant case studies in the influence of political marketing and communication in Canada. Specifically, the market orientation and the need to develop a palatable brand have led these left-wing parties to temper their strict adherence to ideology. Their use of inoculation, moderation, and simplification demonstrates the NDP's acceptance of working within the confines of the generally accepted Canadian political discourse. At the same time, it reflects the parties' attempt to develop the Nike-like superbrand that defines the very politics of Canadian (Manitoban and Nova Scotian) society itself. The extent and longevity of its success will be determined in elections to come.

The extent and longevity of its success remains in question. The precarious situation the Manitoba and Nova Scotia NDP face, and the possibility of the federal NDP slipping back to third-party status, underlines the constant need for political parties to maintain their brand. The challenges faced by the federal and Manitoba NDP are understandable considering the strong linkage between a party's brand and its leader. Both parties underwent leadership changes, with Gary Doer stepping down as Premier of Manitoba in 2009 to become Canada's ambassador to the United States, and the death of Jack Layton in August, 2011. In addition, a preliminary examination suggests that the recent decline of the Manitoba and Nova Scotia NDP coincides with a shift away from moderate, third-way policies. In Manitoba, this has included amending the province's balanced budget legislation and breaking an election promise not to raise the provincial sales tax. Likewise in Nova Scotia, the NDP continued to struggle with inoculating itself against financial/economic issues as Premier Dexter resorted to increasing taxes in order to balance the budget near the end of his mandate. The federal NDP, too, has strayed somewhat from Layton's moderate approach, particularly on issues related to resource development. The ability of each party to maintain (and in some cases re-establish) its brand will be a leading determinant of future success.

6

The Not-So Social Network: The Use of Twitter by Canada's Party Leaders

Tamara A. Small

The Internet is thought to have the capacity to reinvigorate and, perhaps, revolutionize politics by enhancing citizen participation. Unlike older communication technologies like television or radio, the Internet is considered to possess democratic characteristics, including interactivity, disintermediation, and accessibility (Bentivegna 2002). By taking advantage of these characteristics, individuals could advance common political interests and facilitate communication with other citizens and political institutions. As Joe Trippi, campaign manager for Howard Dean's 2004 presidential primary campaign, puts it, the Internet is "the best tool ... ever created" to help achieve full participation in democracy (2004, 226). Indeed, the term "e-democracy" demonstrates how linked democracy and the Internet are in the minds of some.

E-democracy claims have become stronger in the era of social media. Social media applications such as Facebook, YouTube, Twitter, and blogs (see Chapter 12 in this book) are fast becoming ubiquitous. These applications share crucial features:

> Social media is the democratization of information, transforming people from content readers into content publishers. It is the shift from a broadcast mechanism to a many-to-many model, rooted in conversations between anchors, people and peers. Social media uses the "wisdom of the crowds" to connect information in a collaborative manner. (Evans 2008, 33)

Social media is *social* – however tautological that may sound. It is about collaboration, co-production, active contribution, and conversation among people. It is not that these characteristics were non-existent previously.

Rather, these features are now built directly into social media applications. Consider how easy Facebook makes it for people to share information about themselves with others. These social characteristics are also connected to claims about democracy. The controversial 2008 Iranian presidential election and the Arab Spring are often regarded as instances where social media has had a democratizing role. The lesson from events such as these is that, through social media, citizens living under repressive regimes can share and access information, bypass censors, and enhance mobilization efforts (Shirky 2011, 28). Even in stable and democratic regimes, social media can play a role. Rather than being passive consumers of political information, individuals can become active participants by sharing information and publishing opinions on social media sites (Loader and Mercea 2011, 759). Moreover, social media can potentially forge "new institutional structures" in government (Bertot et al. 2010, 53). Essentially, the social capabilities inherent in social media may improve democratic politics by creating a platform for conversation, connectedness, and participation between governors and the governed.

This chapter considers this potential within the Canadian electoral context. Drawing on data from the Twitter feeds of the leaders of Canada's political parties, it asks whether Twitter is used as a *social* media. Established in 2006, Twitter is a fusion of two types of social media: social networking and microblogs. Twitter allows subscribers to write a 140-character update called a tweet. It uses a friendship model whereby clients select which Twitter users to monitor while themselves being followed by other Twitter clients (Marwick and Boyd 2011, 116). The 2011 Canadian election was dubbed the "Twitter Election" by the media. According to the Canadian Press (2011c), "Twitter is the new 'amplifier' for political leaders aiming to mobilize supporters and keep the pressure on opponents." The technology gained prominence as a political tool during Barack Obama's 2008 presidential campaign. In Canada, Twitter also made its debut in 2008, with all five main political parties operating accounts. However, at the time, as in the United States, it played second fiddle to the social media giant Facebook. More recently, Twitter has become the technology du jour. One reason for this is its exponential growth, while the number of Canadians using Facebook has stagnated. In 2009, less than 1 percent of Canadians used Twitter; two years later, that number had grown to almost 20 percent (Ipsos 2011). Twitter's popularity makes it worthy of scholarly attention within the field of political communication.

There is a small but growing literature on politicians' use of Twitter (see Glassman, Straus, and Shogan 2009; Solop 2009; Golbeck, Grimes, and Rogers 2010; Grant, Moon, and Grant 2010; Small 2010a; Ancu 2011; Hendricks and Frye 2011; Lawless 2012; Parmelee and Bichard 2012; Zamora Medina 2012). Despite differing research objectives, one observation is common: most research has found that politicians rarely use the social or conversational aspects of Twitter. In some ways, President Obama has become synonymous with social media campaigning. Although it is true that Obama used social media in innovative ways in 2008, Ancu's (2011) assessment should give us pause. She describes the use of Twitter by Obama and his opponent, John McCain, as "shallow" (20). Overall, McCain engaged minimally with Twitter, tweeting only 28 times in the two months prior to the election. Obama, on the other hand, tweeted 261 times. However, she notes, "The Obama campaign did not produce interactive content" on Twitter, and the feed was strictly one way (Ancu 2011, 16). In other words, Obama broadcasted on Twitter, which is indicative of Web 1.0 content control prevailing over Web 2.0 interactivity. Other American studies have found similar results. Golbeck, Grimes, and Rogers's (2010) study of more than 150 congressional Twitter feeds found that only 7.4 percent of the tweets were considered conversational. This is not only an American phenomenon. Small's (2010a) examination of twenty-seven Canadian political party and leader feeds found that less than 16 percent of tweets were conversations between politicians and followers. In a comparison of politicians and regular users in Australia, Grant, Moon, and Grant (2010) concluded that although politicians were more engaged than Australians on Twitter, this engagement was not related to political dialogue. Finally, Zamora Medina (2012) found that Spanish presidential candidate Alfredo Pérez Rubalcaba made little use of the features that encourage social interaction (retweets and mentions). She concludes that Twitter was a propaganda tool rather than a communication platform in Rubalcaba's campaign. These conclusions, however, could apply to elite political actors in varying political contexts.

Case Study

METHOD
Content analysis is the methodology used in this research. It is a common method in e-research (Anderson and Kanuka 2003, 174) that generates data that are valid, rigorous, reliable, and replicable (Sampert and Trimble 2010).

The focus is on the communication flowing from party leaders to the public during the 2011 Canadian election. The leaders were just five voices among thousands in the Twitterverse, yet they were among the most important political communicators in the campaign. The main unit of analysis is the individual tweet, which has a 140-character limit. However, a lot can be conveyed in 140 characters, including URLs directing readers to other sites, and hashtags that group users' contributions, personal information, policy, reactions, and conversations with other politicians, journalists, and citizens (Parmelee and Bichard 2012, 167). Every tweet from the accounts of the five leaders between the day the writ dropped (March 23, 2011) and Election Day (May 3, 2011) was collected, for a total of 788 tweets over thirty-eight days.[1] The coding scheme is based on research by Small (2010a). Each tweet was designated as one of the categories listed in column 2 of Table 6.1. The tweets were hand-coded by a single coder.[2] Finally, the number of tweets and followers was recorded for every week of the campaign. As Table 6.1 shows, this analysis makes a distinction between social and broadcast tweets. Because social media facilitates two-way conversation, it is believed to be antithetical to broadcasting, which focuses on one-way, top-down communication flows (Mayfield 2008, 5). Indeed, sites like Twitter are "designed primarily to facilitate conversation among individuals and groups" (Kietzmann et al. 2011, 244). Two features, @replies and retweets, are typically used to assess the level of social interaction of political actors (see Grant, Moon, and Grant 2010; Small 2010a; and Parmelee and Bichard 2012). The @replies allow Twitter users to directly and publicly respond to their followers. Honeycutt and Herring's (2009, 9) study on Twitter conversations describe the application as "noisy," (1) where conversations between users can successfully take place. Like email forwarding, a retweet is a reposting of a tweet of another user. Boyd, Golder, and Lotan (2010, 1) argue that retweeting "contributes to a conversational ecology in which conversations are composed of a public interplay of voices ... Retweeting brings new people into a particular thread, inviting them to engage without directly addressing them." Moreover, retweets are social, as they imply listening (Grant, Moon, and Grant 2010, 594). To retweet, one must read the tweets of those one is following. Broadcasting is normally attributed to media organizations, but Lilleker (2006, 46) suggests that broadcasting is any form of communication directed toward mass audiences. It occurs when information flows in one direction from a single sender to the audience. Under this definition, political advertising, press releases, and Internet communication would be included.

Table 6.1

Coding scheme of party leaders' tweets

Social tweets	@reply	Messages sent from one person to another over Twitter; distinguished by the @reply.
	Retweet	The reposting of other people's tweets; distinguished by the formulation "RT @user."
Broadcast tweets	Events	Tweets providing information about future political events.
	Political	Tweets about policy or political issues, including criticism of other parties that is not official party communication.
	Personal	Tweets about matters unrelated to politics.
	Party	Tweets about party- or government-related, activities including policy announcements, press releases, and other documents.
	Reporting news	Tweets about current events and news.
	Status update	Tweets about what one has done, is currently doing, and is going to do.
Other		Tweets that do not fit in any other category.

Here, broadcasting tweets are any that do not attempt to be social. As Table 6.1 shows, broadcast tweets are divided into several politically oriented categories. This methodology allows us to assess how social Canadian party leaders were on Twitter, as well as the types of political information that was communicated during the election campaign.

Results

Twitter, by nature, encourages frequent posting, because each tweet is only 140 characters. Table 6.2 shows that the five politicians tweeted a total of 788 times over the thirty-eight-day campaign. All of the leaders posted at least one tweet for every day of the election, but most tweeted several times per day. One way to contextualize these data is to compare them with the 2008 Canadian federal election (see Small 2010b), when all of the major party leaders had Twitter accounts for the first time. In the 2008 campaign, the five leaders posted a total of 468 tweets.[3] Thus, 68 percent more tweets were produced in 2011. Does this mean that Canadian party leaders and their media handlers see Twitter as a more effective way to communicate

TABLE 6.2

Tweets by leaders during the 2011 federal election campaign

	Party	Account	Total tweets	Tweets-per-day rate	% of tweets by leader
Gilles Duceppe	BQ	@GillesDuceppe	100	2.6	12.7
Stephen Harper	CPC	@pmharper	111	2.9	14.1
Michael Ignatieff	LPC	@M_Ignatieff	73	1.9	9.3
Jack Layton	NDP	@jacklayton	85	2.2	10.8
Elizabeth May	Green	@ElizabethMay	419	11.0	53.2
Total			788	—	100.0

NOTE: The tweets-per-day rate is calculated by dividing the total number of tweets by the total number of days of the writ period (thirty-eight).

than they did previously? This is plausible, but it does not appear likely. Much of the increase in tweets comes from a single feed, @ElizabethMay. In 2008, Elizabeth May had the second-highest number of tweets (behind Jack Layton), with a tweet per day rate of 3.1. In 2011, her rate was 11. Indeed, her tweets comprised over half of all the tweets issued by party leaders, despite her not having comparable human and financial resources (May did not hold a seat in Parliament, and her primary goal was to get elected in her riding). The variation of tweets between the other leaders was minor. Even though 2011 was considered the "Twitter Election" by journalists, except for May, the leaders used the application in the same way that they did in the 2008 campaign in terms of producing content. This is a significant observation given the growth of the medium and its potential for e-democracy.

Political communication is concerned with the transmission of information between politicians, the media, and citizens. For Twitter to be effective for political actors, there must be people who receive their tweets – that is, followers. Otherwise the transmission process is unsuccessful. Followers are the list of people who receive tweets of another user. In their study of Twitter influence, Cha and colleagues (2010, 12) describe the number of followers as a measure of "indegree influence" or popularity, as it "directly indicates the size of the audience for that user." Although indegree influence can tell us much, there are limitations. On the one hand, the number might underestimate the influence. Twitter allows for protected and public tweets. Protected tweets are visible only to approved followers, whereas public tweets are visible to anyone. Because the Twitter feeds of the five leaders are

TABLE 6.3

Indegree influence by leader during the 2011 federal election campaign

	First day of campaign	End of campaign followers	New followers	% of change for new followers
Duceppe	46,097	58,247	12,150	26.4
Harper	103,837	136,143	32,306	31.1
Ignatieff	64,704	98,180	33,476	51.7
Layton	60,796	98,469	37,673	62.0
May	11,955	25,180	13,225	110.6
Total	287,389	416,219	128,830	44.8

public, their tweets may reach more people than just followers. Moreover, tweets reach beyond the Twitterverse, if the content of a tweet is reported in the mainstream media. On the other hand, the number of followers "only indicates how popular a person is, not how many people actually read the posts" (Kietzmann et al. 2011, 247). Table 6.3 shows the indegree influence of the five politicians during the 2011 campaign.

There has been an increasing interest in Canadian party leaders in the Twitterverse. In 2008, just over 4,000 people followed the leader feeds (Small 2010b). The number of followers of the leaders had increased by 101 percent, to just over 400,000 followers, at the conclusion of the 2011 campaign (see column 5 of Table 6.3). Not only has there been considerable growth between the two elections but there was also an increase of almost 45 percent in the number of followers during the 2011 writ period alone. The election clearly generated excitement in the Twitterverse. For thousands of people, it became the place to get daily news about the campaign. More than 38,000 Twitter users joined in week one, and by Election Day, more than 128,000 had started following a party leader for the first time.

Here we take a closer look at the tweets produced by the leaders. Given the anonymous nature of the Internet, there is no way to determine who is actually doing the tweeting – the leader or a staff person. Potentially, there could be differences in how each tweets. Nevertheless, the data do provide insight into how politicians and their campaigns conceive of Twitter as a communication tool. And as will be shown, there are considerable differences in how Twitter was used during this election. Previous research demonstrated politicians seldom use the social aspects of Twitter. Figure 6.1

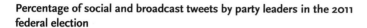

FIGURE 6.1

Percentage of social and broadcast tweets by party leaders in the 2011 federal election

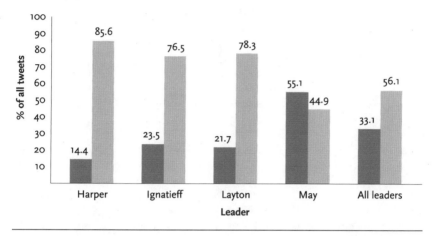

and Table 6.4 summarize the content analysis regarding social tweets. Figure 6.1 shows the percentage of broadcast versus social tweets; Table 6.4 breaks down the social tweets into its components. Overall, the leaders produced drastically more broadcast tweets than social ones (Figure 6.1). Two-thirds of all tweets were categorized as broadcast. This is consistent with previous research. However, the difference between Elizabeth May and her four competitors is notable. Not only did she tweet more, but her tweets were considerably more social.

TABLE 6.4

Total @replies and retweets made by party leaders during the 2011 federal election campaign

	@replies (%)		Retweets		Total tweets
Duceppe	4	(4)	9	(9)	100
Harper	15	(14)	1	(1)	111
Ignatieff	6	(12)	6	(12)	51
Layton	11	(16)	4	(6)	69
May	184	(49)	21	(6)	372
All leaders	220	(31)	41	(6)	703

Almost 50 percent of all May's tweets were @replies, where she responded directly and publicly to followers (see Table 6.2). May engaged socially using @replies with her followers in several ways. First, she answered questions and provided information requested by followers. One example of this is the following exchange between a Green Party member and May:

> *marcperrot:* @elizabethmay What time is your press conference today?
> *ElizabethMay:* @marcperrot all the details are here http://bit.ly/eC5oxJ

May also used Twitter to acknowledge her followers who had tweeted messages of encouragement and well wishing. For instance:

> *ElizabethMay:* Thanks so much! @ryantdavies @ryantdavies @GeorgieBC @PKingI74 @saskgreen @sebastienlabont @SpartanVTyranny @gorgeousgrrly @johannabee

While these acknowledgment tweets may not seem like much, it does show that during the campaign, May read tweets sent to her. The thank-you tweets demonstrate a desire to connect with followers even in a small way. May's "Question of the day" tweets also had a social aura.[4] She posted a question on Twitter, and interested followers could tweet back the answers. The following day, May tweeted the correct answer. For instance:

> *ElizabethMay:* Q of the Day: What bill removed "the duty to act honestly" from the Ethics Code? #GPC #elxn41 #EMayIn
> *ElizabethMay:* Answer: gutting the CEAA never went to env committee. It was stuffed in the 2010 budget bill and only finance committees looked at it #elxn41

May posed nineteen questions of the day during the campaign. As with the example above, the questions were all related to substantive policy issues such as the Kyoto Accord and were generally critical of the Conservatives. One of May's catchphrases has been "doing politics differently"; this is certainly true of her Twitter feed. For May, Twitter was a social media. There is empirical precedence for this finding. Earlier studies show that the online presences of Green parties worldwide are more advanced and interactive than are other parties (see Ward, Gibson, and Nixon 2003; Small 2008b). Given their precarious electoral position, Green parties have higher incentives to use the Internet and to use it effectively (Ward, Gibson, and Nixon 2003). Social media might prove useful for minor political parties with limited funds and organizations; however, more research is needed to assess this claim.

In this study, Green leader May considerably skews the overall data relating to social interaction. Social tweets played a far smaller role for the other leaders; @replies constituted less than one in five of their tweets. All four leaders sporadically answered questions posed by followers. For instance, Michael Ignatieff answered four questions, most related to his party's Family Pack policy. Bloc Québécois leader Gilles Duceppe used @replies to communicate with journalists, including anglophones outside Quebec. Perhaps one of the most intriguing social moments in the campaign was the tweeting between Michael Ignatieff, Stephen Harper, Gilles Duceppe, and comedian Rick Mercer about a possible one-on-one televised debate between Ignatieff and Harper:

> *pmharper:* @M_Ignatieff curiously, my team proposed 1:1 to TV consortium today; however, your team did not speak up.
> *M_Ignatieff:* @pmharper A one-on-one debate? Any time. Any place.

Duceppe tweeted that Harper and Ignatieff were excluding Quebec from the proposed debate. Mercer then tweeted an offer to moderate a Harper-Ignatieff debate – and $50,000 for the charity of their choice if they participated. Only Ignatieff responded, tweeting "I'm in." Although this exchange lasted only a few days and a few tweets, it received considerable attention from the mass media and was retweeted extensively. However, this type of exchange between the leaders on Twitter, in the form of @replies or mentions, was the exception, not the rule. Twitter has not yet become an alternative venue for political debate among the leaders.

Retweeting can also inform our understanding of social interaction. Retweeting embodies many social media characteristics, including listening, reciprocity, and co-production. It implies the reading of the tweets of others. Moreover, it implies the amplification and validation of those you

TABLE 6.5

Types of broadcast tweets made by party leaders during the 2011 federal election campaign

Leader		Events (%)	Party (%)	Political (%)	Personal (%)	Status update (%)
Duceppe	(n = 86)	3.5	11.6	66.3	1.2	17.4
Harper	(n = 94)	11.7	78.7	3.2	5.3	1.1
Ignatieff	(n = 57)	12.3	12.3	29.8	7.0	38.6
Layton	(n = 69)	43.5	17.4	14.5	1.4	23.2
May	(n = 150)	14.7	8.0	40.0	3.3	34.0
All	(n = 453)	16.1	25.4	32.5	3.5	22.5

follow. However, as Table 6.4 shows, retweets are uncommon. Only 5 percent of all leader tweets were coded as a retweet. Even for Elizabeth May, who was talkative on Twitter, retweets made up only a small proportion of her posts (5.6 percent). This finding is comparable with previous studies. Golbeck, Grimes, and Rogers (2010) found retweets by US congresspeople to be infinitesimal; of the 4,626 tweets examined, only five were retweets. Listening and reciprocity do not appear to be important goals. This provides greater evidence of the general disregard for social interaction by Canadian politicians.

So within Canadian campaign politics, Twitter is the "not-so social" network. For most of the leaders, Twitter is treated as a broadcast channel. If Twitter is not being used for social interaction, what are the leaders broadcasting? Table 6.5 breaks down the broadcast tweets into political categories. Overall, no one particular broadcasting style can be applied to all leaders during the election campaign. Rather, each leader had his or her own particular tweeting style in terms of content.

Bloc Québécois leader Duceppe and Green leader May (when she was not being social) can be described as "political" twitterers. As an example:

> *GillesDuceppe:* Harper lied to the people, he agreed with the coalition http://bit.ly/fIeItv [translated by author]
> *ElizabethMay:* So Harper has borrowed part of #GPC income splitting policy, but partial and only after deficit eliminated. We use tax shifting to do it now.

Political tweets are tweets about policy or political issues, and include criticisms of other parties that are not official party communication. In the first example, Duceppe tweets that there was a plan to build a Conservative-NDP-Bloc coalition in 2004 despite Harper's denial of any such plan. The tweet provided a link to a copy of a letter sent to the governor general that mentions the three party leaders, Harper among them. In the second example, May responds to a Conservative campaign announcement. As we can see, political tweets reflect what a politician is thinking at a given moment and are generally in reaction to campaign events. In this sense, they are individualized. Political tweets are topical, timely, and less managed in their format. Although Layton and Harper used this style of tweeting nominally, political tweeting was the most common type of broadcasting in this campaign, with nearly a third (32.5 percent) of all leaders' tweets dealing with such political matter.

Party tweets were the second most common type of broadcast tweet used by the party leaders. This type of tweet is focused on official party communication. They typically refer to policy announcements, press releases, and other party documents, usually combined with a URL. For example:

> *pmharper:* Today, I announced we will help families save $1,300 per year, on average, by sharing their income for tax purposes. http://is.gd/YW5iZp

The URL links to news releases posted on the party website. As the example shows, party tweets are tied very much to one goal of politicians: to stay on message. Whereas political tweets have a spontaneous feel, party tweets are deliberate. Moreover, party tweets mean that Twitter varies little from the official party website.

Perhaps it is not surprising that the main proponent of this style is the prime minister. Almost 80 percent of Harper's tweets were coded as "party." Harper has been both lauded and criticized for his strict communication style. Former Conservative campaign manager Tom Flanagan (2007, 283) argues that a disciplined communication strategy is central for Conservatives; accordingly, the leader should never improvise, candidates should never speak about personal beliefs, and electronic communication must be "careful and dignified" (284). These rules are evident in Harper's tweeting. Harper rarely tweeted spontaneously, and his tweeting was focused on the communication goals of the party. Indeed, Harper, as an individual, was quite absent from his tweets, which was a different style from that of his competitors. Party tweets represented an average of only 17 percent of all broadcast tweets for the other leaders.

Besides a disciplined communication style, there is another reason for this difference between Harper and the other leaders, one related to an earlier research finding. Small (2010a) found that a political party may sometimes operate more than one Twitter feed, each having its own communication style. Except for the Conservatives, the others operated feeds for the national party that were distinct from the leader feed (e.g., @NDP_ElectionHQ, @CanadianGreens, @liberal_party, and @BlocQuebecois). These feeds focused mostly on official party communication, which left the leader feeds to focus on leader-oriented topics such as political tweets and status updates (ibid.).

In some ways, Layton's tweeting style was similar to Harper's in that the party played a greater role in tweets. Almost half of Layton's tweets were focused on promoting future campaign events, especially online ones. For instance:

> *jacklayton:* Online or in person: join me for a huge rally in St. John's today – with live music by Idlers http://ndp.ca/hhuNZ #NDP #Cdnpoli #elxn41

As part of their tours, party leaders criss-cross the country, going from event to event. One goal of the tour is to give voters a chance to see a leader first-hand (Taras 1990, 154) and to generate local media coverage. Given the size of the country, this goal is difficult to achieve. However, the NDP has been using technology to make it a reality. The NDP live-streamed the majority of its events on its website and used Twitter to encourage online attendance. Encouraging techno-savvy followers to get involved in the campaign can be seen as an effective mobilization strategy. Arguably, Twitter followers would be the very type of people to be mobilized, as they have already indicated an interest in the party. As such, it is surprising how uncommon this tactic was during the campaign. Less than 20 percent of all leader broadcast tweets informed about and encouraged attendance at upcoming events.

In an earlier study, status updates were the most common type of tweets by Canadian federal and provincial party leaders: 54 percent of leader tweets were of that nature (Small 2010a). In this style, tweets focus on what the leader did or is going to do, or where the leader has been or is going to be.

Status updates give followers an opportunity to see what politicians do on a daily basis, including travelling, and meeting with journalists and the public. Status updates played a far smaller role in the 2011 campaign, making up only 22.5 percent of all broadcast tweets. It was, however, the main tweeting style of Liberal leader Ignatieff. Almost 40 percent of his tweets were similar to this one:

> ▸ *M_Ignatieff:* Driving to Montreal after rally in Quebec City. Inspired by last night's rally with Jean Chrétien. Watch it here: http://lpc.ca/rur #elxn41

If Harper was very much absent from his tweets, Ignatieff's tweets were heavily individualized. Followers of his tweets would get a good sense of what life is like on the leader's tour. However, of the categories of broadcast tweets that have been discussed, status updates are the least politically relevant. Party tweets may lack spontaneity, but they do provide political information for voters. The same can be said for political tweets in that they provide insight into the thoughts and reactions to campaign events and current political issues by a politician. Event tweets are also politically relevant by mobilizing supporters. Status updates, such as "Great response at the gates to UVic! Waving at the morning commuters. #elxn41@MyUVic #GPC" may be interesting but cannot be said to aid followers' campaign learning.

Overall, the data indicate that, with the exception of the Green leader, Canada's top federal politicians did not see Twitter as a social media. Generally, this finding is consistent with previous research. However, May is a significant outlier in this study, and perhaps within the literature. There does not appear to be only one Canadian way to use Twitter. Rather, different politicians make different choices about how to use it. Elizabeth May chose to be social, whereas the other leaders did not. However, even among the others, there was no single way to broadcast in this campaign.

Political Communication in Canada

How do we understand these findings? What does it mean that the leaders of Canadian federal parties do not use Twitter for social purposes? Given that many consider social media to be antithetical to broadcasting, should we chastise Canadian politicians for their incorrect and perhaps even anti-democratic use of Twitter? For instance, *ComputerWorld* (2011) noted after the election, "In fact, it was everyone but the politicians who managed to leverage social media successfully during the campaign."

This finding is consistent with previous research on Canadian political parties and Internet interactivity. In 2005, Darin Barney wrote that Canadian parties "have been very reluctant to pursue with vigour and creativity the potentials that ICTs [information and communication technologies] present for the meditation of more routine, deliberative, participatory exercises" (140). Various studies on elections since 2004 show very limited use of interactive features by Canadian parties (see Kernaghan 2007; Small 2008b, 2010b). Instead, Canadian parties have used the Internet to supplement their traditional campaign activities, including for media relations, fundraising, and organizing. It is not that the Internet isn't having an impact on Canadian party politics; rather, research shows that its impact on e-democracy appears minimal, regardless of what journalists claim during a campaign. And despite the introduction of new social media such as Twitter, this study shows that Barney's statement still has salience.

Second, for those who are interested in political communication research, the focus on the conversation as the sole function of social media campaigning limits the understanding of social media and the role it can play. Carpentier (2011, 200) points out that, when discussing new media, "we are often led to believe that all audiences of participatory media are active participants, and that passive consumption is either absent or regrettable." However, passive consumption of Twitter is more common than one might think. For instance, research conducted by Sysomos Inc. (2010) found that 2.2 percent of users account for about 60 percent of all tweets. Finally, Twitter (2011) has found that 40 percent of active users never tweet at all.

What this means is that a small number of accounts are communicating with a mass audience. Twitter is not solely about tweeting; it is also about listening. These numbers show that many users are not interested in posting tweets, nor are they interested in engaging in conversation or even retweeting what they have read. For many users, Twitter is not so much a social network as it is a broadcasting channel. It is more like Wikipedia than Facebook: many people read but do not actively contribute. Twitter is a place to aggregate and streamline content from many sources of interest to a single place.

So despite the main finding of this research being similar to that of previous Canadian studies, we should not necessarily chastise Canadian politicians and their communications staff for how they used Twitter in the 2011 election campaign. In terms of political communication, broadcasting tweets can be seen as an effective use of Twitter, because many people appear to want to hear what leaders have to say. Indeed, there might even be potential benefits and incentives for broadcast tweets. Twitter provides an unmediated opportunity to connect with a very desirable group – the mainstream media. According to Taras and Waddell (2012b, 97), Twitter has replaced the BlackBerry for direct messaging between the politicians and journalists. The ability of politicians to send their latest political musings or press releases directly to the media through a tweet is an efficient media relations strategy. A second desirable group of Twitter followers are the politically engaged. American research by the Pew Research Center's Internet and American Life Project shows that individuals who use social networking sites as an outlet for political engagement are far more active in traditional realms of political participation than other Internet users; their activities include contacting the government, joining an interest group, and expressing themselves in the mass media (Smith et al. 2009, 17; no comparable Canadian data exist). This is an important group of people that politicians want to influence, get the message out to, and encourage mobilization of, because these people are likely to vote. Finally, broadcast tweets can be a source of campaign learning for voters. As we saw earlier, both party and political tweets provide voters with the perspectives of the leader and the party. By using URLs, the 140-character tweets can provide considerable information about campaign promises and leader views. This said, the usefulness of Twitter as a broadcast medium does in part depend on the types of information tweeted. As noted, political and party tweets can provide useful political information to citizens and the media. Event tweets, like Jack Layton's, will be of interest to those followers looking to get more involved in the campaign. The relevance of status updates is minimal.

There appears to be strategic benefits of broadcast tweets in targeting individuals who follow politics closely. However, this highlights a broader issue about Internet politics: the democratic divide. According to Norris (2001), the democratic divide refers to the differences between those who do and who do not use digital technologies for participation and mobilization. The democratic divide has the potential to exacerbate political divides (income, educational, racial) that already exist. The very targeted nature of social media technologies such as Twitter may contribute to a democratic divide by creating more opportunities for the politically engaged to access politicians and participate in politics.

Nevertheless, these potential benefits require Canadian research. The literature on how Canadian parties use the Internet, including such social media, is rich and detailed, but more research is needed that explores how citizens and journalists engage with that broader world of Canadian party politics online.

NOTES

1   My thanks to Kevin Geiger for his research assistance in collecting the data.

2   There are limitations to using a single coder. As Heck (2004) notes, a single coder may code items consistently, but there are concerns of reliability.

3   With the exception of the Liberal Party of Canada, which was led by Stéphane Dion in 2008 and by Michael Ignatieff in 2011, all the parties were led by the same people in 2008 and 2011.

4   The Q and A tweets were not coded as conversational, nor were @replies or re-tweets. These tweets were coded as "other" (neither broadcast nor conversational).

PART, 2

Canadian Political News Media...............

7

The Canadian Parliamentary Press Gallery: Still Relevant or Relic of Another Time?

Daniel J. Paré and Susan Delacourt

Chapter 6 suggests that the ability of politicians and citizens to connect without the filter of the mass media is changing the nature of political communication. So what does the Internet age mean for traditional political media? The journalists who comprise legislative press corps and galleries are commonly understood as playing a particularly important role in the functioning of liberal democracies given their position as the primary intermediaries between citizens and their governments. Ideally, they are meant to act as political watchdogs, monitoring for abuses of power and disseminating the information citizens need to participate effectively in the political process. The extent to which contemporary political journalism lives up to these lofty goals is a matter of debate, yet a free and vigilant press publicly reporting on affairs of the state is nonetheless widely considered a pillar of democracy and essential to good government. In the words of one of the most respected North American newspersons of the twentieth century:

> A democracy ceases to be a democracy if its citizens do not participate in its governance. To participate intelligently, they must know what their government has done, is doing and plans to do in their name. Whenever any hindrance, whatever its name, is placed in the way of this information, a democracy is weakened and its future endangered. This is the meaning of freedom of the press. It is not just important to democracy; it is democracy. (Cronkite 1984)

Within parliamentary systems, groups of journalists accredited to cover the activities of Parliament are usually referred to as a press gallery. The equivalent organization in the United States is the White House Correspondents' Association. Historically, these organizations have been tasked with surveying socio-political developments, identifying the most pertinent issues of the day, providing platforms for deliberation and debate, holding public officials to account, incentivizing civic engagement, and resisting the subversion of media independence (Seymour-Ure 1962; Fletcher 1981; Gurevitch and Blumler 1990; Riddell 1998; Ritchie 2005).

For many Western democracies, the organization of political communication throughout the last century advanced through three successive, and at times overlapping, stages (Levine 1993; Schudson 1998; Blumler and Kavanagh 1999). The first, which in Canada spanned from the mid-1800s to the early 1960s, was marked by a largely print-based media system that was organized along partisan lines, and which emphasized policies and policy directions in political news reporting. The second phase corresponded with television becoming the dominant mass media platform, a concomitant increase in audience size, the adoption of a more confrontational approach to political reporting, and waning party loyalty of voters. The third, or present, phase combines abundant flows of digitally mediated political information, a shift toward news reporting decision making that is largely driven by commercial considerations, the entrenchment of adversarial journalism, the proliferation of news management and other strategic communication techniques in the political domain, and reductions in civic engagement (Bennett and Manheim 2001; Pfetsch 2007; Taras and Waddell 2012a).

The dynamics shaping the contemporary contest for the news agenda and public opinion involve fluctuations in power relations between media and government institutions that call into question both the traditional gatekeeping functions of the mass media and its agenda-setting role (Bardoel 1996; Dahlgren 2009). Drawing from Shoemaker and Vos (2009, 1), we define gatekeeping as the process by which social reality pertaining to political developments is constructed and disseminated by the mass media through the "process of culling and crafting countless bits of information into the limited number of messages that reach people each day" (for more on gatekeeping, see Chapter 10 in this book). This process determines the information selected for dissemination, as well as the content and framing of the transmitted messages. It also has direct implications for how, and the extent to which, the mass media carry out the democratic functions they are expected to serve.

Gans (1979) famously likened the relationship between journalists and their political sources to a tango in which both parties attempt to leverage the resources they control in order to lead the dance. The contest pits the power of political news sources to control information and to grant legitimacy to news stories against the power of journalists to control the content and framing of news stories. Whereas politicians seek to ensure "the delivery to voters of as full and unchallenged versions of their messages as possible," journalists are equally vigilant about upholding their right to independently analyze, interpret, and comment on political developments as they see fit (Blumler and Gurevitch 1995, 205). For much of the last quarter century, the common view within the Canadian context has been that it was journalists who were leading the dance (Taras 1990; Nesbitt-Larking 2001; Flanagan 2007). In other words, although political strategists may set the agenda, they too have needed to play by the rules of media logic (Fletcher and Everett 2000).

Adopting a systems perspective, Blumler and Gurevitch (1995) argue that changes in the relations between political and media institutions are conditioned by interactions among four sources of instability. The first two variables relate to the broader environment within which political and media institutions are situated. They are (1) innovations in the technologies used to produce and diffuse political messages, which, in turn, impact on the structuring of messages, audiences, and democratic engagement (see, for example, Coleman and Blumler 2009; Papacharissi 2010), and (2) alterations in the socio-cultural and socio-economic context within which a political communication system is situated (see, for example, Hallin and Mancini 2004; Ward and Gibson 2009). The other two sources of instability are internal to the relations political and media institutions share, and pertain to (3) the motivations and drives of journalists and politicians to continually adjust their activities in response to one another's strategic manoeuvring (see, for example, Entman 2004, 2007; Marland 2012), and (4) the need to adjust the structure, style, and tone of political communications so as to maximize audience patronage in light of political objectives and commercial pressures (see, for example, Bennett and Manheim 2001; Strömbäck 2011; Alboim 2012).

The recognition that changes in political communication systems are instantiated in highly dynamic, fluid environments underpins the model of news construction advanced by Bennett (2004). He identifies four variables that he maintains simultaneously influence how political news narratives are selected and constructed, and which he associates with four ideal-type gatekeeping constructs. The first variable is the news sense of individual journalists. It is linked to a journalist-driven construct of gate-keeping that corresponds with the romanticized notion of journalists as independent gatekeepers serving an engaged citizenry as investigative watch-dogs over government officials, with news-making decisions driven by tacit knowledge and implicit values. The second variable is the routines and stan-dards of the organization with whom the journalist is employed. Organ-izational gatekeeping, the second ideal-type, is rooted in the bureaucratic routines guiding the activities and decisions of journalists employed by news organizations. The role of journalists, here, is that of authoritative record keeper.

The third variable is the economic constraints on news production. Here, emphasis is placed on how the economic and political interests of a parent corporation influence decision making within the news organiza-tions that it owns (see, for example, Herman and Chomsky 1988; McChesney 1999; Taras 2001). Concerns about the latter are linked to reductions in in-formation diversity resulting from the concentration of media ownership. The journalist's role within the context of economic gatekeeping primarily entails finding content to fit into information formats that seek to deliver the most profitable audiences to advertisers, including the use of routine news stories as opportunities for punditry. The fourth variable is the information and communication technologies that facilitate the production and diffu-sion of information. The notion of technology-driven gatekeeping is prem-ised in the immediacy of digitally enabled information flows, with the journalist's foremost role being that of information transmitter and the pub-lic treated as a voyeur.

During any given period, one or more of the above variables may dominate how the gatekeeping process is manifest, resulting in periods of equilibrium that come to constitute "normal" working conditions. The equi-librium is upset when the dominance of one or more of the variables is dramatically altered.

Within Canada, the dominant system of gatekeeping that has histor-ically been in place at the national level is a journalist-driven/economic hybrid involving the government of the day and the Canadian Parliamentary Press Gallery (CPPG). The history of press-government relations in this country has been marked by fluctuations between periods of healthy antag-onism between politicians and journalists and outright hostility as both groups of actors vie for control over the political news received by the public and how that news is framed. Levine (1993), for example, traces discord between the Prime Minister's Office and the CPPG back to Prime Minister John A. Macdonald's (1867-73, 1878-91) efforts to control how he and his party were portrayed in the press. Relations between the CPPG and Richard Bennett (1930-35) were particularly caustic, with journalists from liberal and conservative newspapers rallying against him. William Lyon Mackenzie King (1921-26, 1926-30, 1935-48) was deeply suspicious of the press, said as little as possible to journalists, and restricted their access to him. Both Louis St. Laurent (1948-57) and John Diefenbaker (1957-63) initially enjoyed positive relations with the CPPG that soured over time. In the mid-1960s, Lester B. Pearson (1963-68) suspended cabinet scrums in response to the increasingly confrontational style of journalism being practised. His successor, Pierre Trudeau (1968-79, 1980-84), was renowned for his disdain of the press gallery. He actively sought to control the lines of communication with the public by limiting access to himself, favouring particular journalists for interviews, and releasing announcements to suit the agenda of his government. In 1976, he replaced media scrums with weekly press conferences, which, much to the chagrin of the CPPG, he regularly avoided. Joe Clark (1979-80), John Turner (1984), and Brian Mulroney (1984-93) all maintained that their administrations had been gravely damaged by the CPPG. Likewise, Jean Chrétien's (1993-2003) and Paul Martin's (2003-6) relations with the CPPG were far from harmonious.

Shortly after the Conservative Party of Canada won the January 2006 federal election, relations between the two dance partners essentially imploded in the light of the PMO's implementation of a communication strategy that, in many ways, resembles the system in place in the United States to guide relations between the White House and the White House Correspondents' Association (Bohan 2012), and at Westminster in the United Kingdom (Parker 2012).

The outcome of this confrontation, combined with Stephen Harper's public acknowledgment of his desire to circumvent the CPPG's position as principal interlocutor of federal politics and his party's use of alternative communication channels (e.g., advertising, podcasts, direct emails, and relationships with sympathetic radio talk shows, newspapers, and conservative bloggers) (Libin 2006), raises questions about the continued relevance of national media organizations like the CPPG. Specifically, is the turbulent press relationship between the PMO and the CPPG reflective of a fundamental shift of the gatekeeping system away from the journalist-driven/

economic hybrid that has long been dominant within the Canadian context? It is to this issue that our attention now turns.

Case Study

This case study focuses on the struggle that erupted between the PMO and the CPPG in March 2006 over the PMO's imposition of new rules of engagement between the federal government and the media. The fracas spanned most of 2006 and some eight years later continues to influence press-government relations. The chronicle of events presented below has been compiled using the minutes of five meetings of the CPPG that took place between April 2006 and May 2012, private email correspondence about the struggle consisting of eighty email messages between bureau chiefs and the editors-in-chief of a number of news outlets that were CPPG members in 2006-7,[1] and the personal notes of one of the authors who was an active participant in the events detailed below. This material, which represents a convenience sample, is supplemented with published media reports and a limited number of available scholarly works.

Founded in 1867 as a self-governing body functioning under the authority of the Speaker of the House, the CPPG views itself as an integral part of the democratic institution of Parliament. In 2006, it consisted of 362 members (including freelancers) representing seventy-four media outlets (see Table 7.1). At the time of writing, it has 379 members (including freelancers) representing sixty-five media outlets. Permanent and temporary membership in the CPPG is open to "journalists, photographers, camerapersons, soundpersons, and other professionals whose principal occupation is reporting, interpreting or editing parliamentary or federal government news, and who are assigned to Ottawa on a continuing basis by one or more newspapers, radio or television stations or systems, major recognized news services or magazines which regularly publish or broadcast news of Canadian Parliament and Government affairs and who require the use of Gallery facilities to fulfil their functions" (CPPG 1987, Section 4).

Membership provides access to areas of Parliament that are off-limits to the general public; special seating within the Senate, Commons, and parliamentary committee rooms; access to the parliamentary reading room; borrowing privileges from the parliamentary library; access to parking space on the Hill; and entry to the annual press-gallery dinners and other, periodic social events (CPPG n.d.).

TABLE 7.1

Canadian Parliamentary Press Gallery membership, 2006 and 2012

Media outlet	Members March 2006	August 2012	Media outlet	Members March 2006	August 2012
Aboriginal Peoples TV	3	6	*Hill Times*	5	9
Agence France-Presse	1	1	Huffington Post Canada		1
ARC Publications	1	1	iPolitics		6
Association de la presse			ITAR-TASS	1	1
francophone	1		La Presse	3	4
/A\ National News		1	La Presse canadienne	2	3
Beta News Agency	1	1	*Lawyers Weekly*	1	1
Bloomberg News	3	5	*Le Devoir*	4	3
Broadcast News	1		*Le Droit*	2	2
Calgary Herald	1		*Le Journal de Montréal*	1	
Canadian Catholic News	1	1	*Le Journal de Québec*		1
Canadian Press	18	19	*Le Soleil*	2	1
CanWest News Service	11		Lobby Monitor		1
CBC News Network	6	10	*Maclean's*	2	6
CBC Radio	12	12	*Medical Post*	1	1
CBC Radio, Network			*Ming Pao News*	1	
Current Affairs	2	1	*Montreal Gazette*	1	
CBC TV	57	52	Muscovite	1	
CCH Canadian Limited	1		*National Post*		1
Central News Agency	1		Need to Know News		2
Christian Current Ottawa	1		Neue Zürcher Zeitung	1	1
CHUM TV	7		*New Brunswick Telegraph-*		
CJAD/CFRB	1		*Journal*	1	
CKOI		1	New Tang Dynasty TV	3	4
CPAC	24	21	OMNI	3	3
CTV	27	32	Ottawa Bureau Inc.	1	
Diplomat Magazine	1	1	*Ottawa Citizen*	11	3
Dow Jones Canada Inc.	1	2	*People's Daily of China*	1	1
Economist		1	Pigiste	1	1
Embassy Newspaper	1	4	PoliticsWatch.com	1	
Empire News		1	Postmedia News Service		17
Epoch Times	1	1	Publinet	2	
Financial Post	1	1	QMI-Quebecor		1
Freelance	18	17	Rabble.ca		1
Global TV	11	17	Radio Canada International	2	1
Globe and Mail	16	16	Radio-Canada Radio	6	4
Halifax Chronicle Herald	1	1	Radio-Canada TV	27	19

▶

◄ Table 7.1

Media outlet	Members		Media outlet	Members	
	March 2006	August 2012		March 2006	August 2012
Réseau francophone d'Amérique	1		TV Ontario	4	
Réseau NTR	1		UPI/Hicks Media	1	
Reuters	5	4	*Vancouver Sun*	1	1
Rogers Broadcasting		1	Vietnam News Agency		2
Science Bulletin	1		Western Producer	1	1
Sun Media	6	19	*Winnipeg Free Press*	1	1
Sunday Edition	1		Wire Report		3
TFO		2	World Business Press Online		2
Times & Transcript	1		Xinhua News Agency	2	2
Toronto Star	11	8			
TVA	6	8	TOTAL	362	379

SOURCE: Adapted from CPPG (2013).

Stephen Harper has been haunted by his portrayal in the media since his days as leader of the Canadian Alliance, and his antipathy toward the CPPG was well known when he became prime minister in 2006 (L. Martin 2006, 2010). Acrimony between the PMO and the CPPG ignited shortly after his new minority government was sworn into office. Since the 1970s, members of the press gallery regularly staked out the corridors outside cabinet meetings to scrum with ministers and to cover photo opportunities with the prime minster. However, in March 2006, the PMO announced its intention to move the cabinet scrums to the lower level of the Centre Block, effectively impeding journalists' access to members of the governing party (St. Martin 2006). This development was followed, a few days later, by the publication of the details of a seven-point email message to all government officials outlining the new government's communication strategy. In addition to mandating all government officials to "maintain a relentless focus" on the Conservative Party's core campaign promises, the message stipulated that the "PMO will have final approval for all communications products – even Notes to Editors or Letters to the Editor," and that ministers were "not allowed to speculate on the future direction of government" (Clark 2006).

Shortly thereafter, changes to prime ministerial press conference procedures were announced. Henceforth, journalists would be required to submit their names in advance to a PMO official charged with deciding who would get to ask a question and the order in which questions would be posed. Previously, it was journalists themselves who decided, either informally or formally, the order of questions. This initiative, dubbed "the list," set off the confrontation in earnest between the CPPG and the PMO, with journalists fearing that the new procedures would be used to favour individuals who were seen to be more sympathetic to the new government and to exclude those whose questions the PMO did not appreciate.

At the end of March 2006, the prime minister's then communications director, Sandra Buckler, met with the CPPG Executive to clarify the new rules and to address complaints about the limited access to cabinet ministers. At this meeting she made clear that there would be little movement on the government's position, noting, "As you probably have noticed, we're a different kind of government and we place a heavy value on communications ... We will retain the option on where we think we best can deliver our message" (Politics Watch 2006).

In addition to registering reservations about using the National Press Theatre for prime ministerial press conferences, Buckler stressed that the PMO would be insisting on the use of "the list," as well as moving forward with limiting journalists' access to the corridor outside cabinet meetings. The transcripts of that meeting, which lasted for approximately twenty minutes, were released to CPPG members, accompanied with a message from then CPPG president Emmanuelle Latraverse, in which she expressed her dismay at "the tone on the part of the PM's staff which left no room for negotiations" (Zerbisias 2006).

At the CPPG annual general meeting on April 7, 2006, relations with the PMO were front and centre, with two resolutions being unanimously passed that constituted a mandate to boycott submitting names to "the list":

1   Be it resolved that the Canadian Parliamentary Press Gallery deplores the Prime Minister's Office's restriction of access, such as access to the third floor outside cabinet and instructs the Executive to take all steps possible to restore the Gallery's traditional access in the spirit of transparent and accountable government (CPPG 2006a, 6).

2   During availabilities by the prime minister and ministers, the Canadian Parliamentary Press Gallery resolves that we will ask questions in the manner that we see fit and will not accept direction from the Prime Minister's Office or anyone who is not a member of the Press Gallery (ibid., 8).

By the end of April 2006, relations between the two parties were deteriorating into public shows of hostility. First there was a boisterous standoff between the prime minister and the CBC's Julie van Dusen over whose turn it was to ask a question during a press conference (Galloway 2006; L. Martin 2010). This was followed, a few weeks later, by some two dozen journalists walking out on a prime ministerial press conference after Harper refused to take questions from members of the CPPG given their boycott of "the list" (Sallot 2006).

In a late May 2006 interview with A-Channel in London, Ontario, Harper decried the CPPG's bias against his government and declared that he would no longer be dealing with them. In his words,

> Unfortunately the press gallery has taken the view they are going to be the opposition to the government. They don't ask questions at my press conferences now. We'll just take the message out on the road. There's lots of media who do want to ask questions and hear what the government is doing. (CBC News 2006)

A few weeks later, Harper further clarified his government's position in an interview with the *Western Standard* (Libin 2006). The gallery, he declared, was run by "left-wing ideologues" and free thinkers who sought to impede him and his party from getting their message out. According to Harper, the CPPG had become "too institutionalized, and too convinced that it can control the news," and therefore, "break[ing] that up in any way" would be "helpful for democracy." He went on to claim that the boycott was actually making his life easier because it enabled him "to pick my interviews when and where I want to have them," adding, "they say if I don't do it their way, I'll somehow gain more control over my media relations. Well, I've got more control now."

Both the newly elected president of the CPPG, Yves Malo, and the then editor-in-chief of the *Globe and Mail*, Edward Greenspon, publicly deplored the PMO's obstinacy regarding the controlling of questioners and expressed a willingness to work toward formalizing conference procedures that would eliminate the chaos traditionally associated with media scrums (Sallot 2006). In a letter sent to the CPPG membership in the first week of June 2006, Malo affirmed the resolve of Canadian media organizations to continue their boycott of "the list." Throughout that summer, a small group of press gallery members, primarily from the Canadian Press wire service, the *Globe and*

Mail, and the *Toronto Star*, began working on a compromise for conducting more orderly news conferences.

In early September 2006, the CPPG convened a special meeting to discuss the boycott, which was now entering its fifth month. In addition to the PMO's failure to return Malo's telephone calls or acknowledge his letters, support for the boycott among the membership was unravelling. Throughout the summer months, CanWest had acquiesced to the PMO's demands, and other media organizations were now considering doing the same. Expressing frustration with the failure of the boycott, one attendee asserted that "the remedy has become worse than the ill it seeks to cure ... The Gallery should accept the list and if there are people not on the list, we can publicly ask the PM why certain reporters are being passed over" (CPPG 2006b, 3). Echoing this view, another participant declared that since the gallery had lost this battle and was now stuck with the list, the real challenge was to determine how to proceed from that point onward.

In a statement reflective of the views of many of those who remained steadfast in their opposition to the list, another attendee proclaimed that

> if the Gallery accepts the list from the governing party it will set a precedent. The other [political] parties will soon have their lists[,] as well as the Canadian Labour Congress, the Chamber of Commerce etc. We're fighting for how people have done our jobs for as long as people have done our jobs and how they're going to do them in the future. Those who have decided to go back on the list for their own reasons will be judged by their peers. (CPPG 2006b, 4-5)

The meeting concluded with the passing of a resolution (fifty-eight for, thirty against, two abstentions) that suspended the "boycott of the PMO's list as a gesture of good faith for 30 days" and mandated the Executive "to talk with the PMO to develop a protocol for press conferences" (CPPG 2006b, 9).

By October 2006, the principled opposition to the list had effectively given way to the commercial considerations of the media outlets that employed individual gallery members. At the end of the month, the CPPG met again to discuss its next steps. It was clear that the PMO was, for the time being, successfully leading the tango insofar as the Canadian public was seemingly indifferent to the Conservative's avoidance of "a cabal of big shot national reporters sitting in Ottawa" (CPPG 2006b). In an effort to find a way around the ongoing impasse, the CPPG resolved to create its own list that would be managed by the gallery. This would at least have the merit of exposing the PMO if its representative skipped over a journalist during press briefings and conferences.

Despite the overtures of the CPPG, the PMO's position remained immutable. At the gallery's 2007 annual general meeting, efforts at proposing an alternative to the PMO's stipulations were effectively abandoned. With no agreement on rules of engagement, dealings between the federal government and the CPPG became increasingly ad hoc. The prime minister held a couple of news conferences in the National Press Theatre in 2007, but his encounters with the gallery were largely contained to occasions when foreign leaders visited the capital.

The Harper government's efforts at leading the tango were also evident in the 2008 and 2011 federal election campaigns. In 2008, Conservative media personnel enforced strict limits on access to Prime Minister Harper throughout the campaign tour, with journalists required to sign lists to ask questions or even gain entry to events (Brennan, MacCharles, and Smith 2008). Controversy erupted a few times over whether the prime minister was using his RCMP security detail to keep reporters in check (CBC News 2008). In the 2011 election, the situation changed little. Stephen Harper agreed to daily encounters with the national media, but these meetings were restricted to five questions – two from anglophone journalists, two from francophone journalists, and one from a local journalist (Chase, Baluja, and Taber 2011). Throughout the campaign, he used interviews with select local, smaller, targeted, and less critical media outlets to communicate with the electorate. Although this was an irritant to the major national news outlets, it did not appear to resonate particularly strongly with voters, who ultimately provided the Conservatives with a mandate to form a majority government.

In May 2012, Andrew MacDougall, Harper's new communications director, requested a meeting with the CPPG Executive (Ryckewaert 2012). This meeting was far more cordial than that which had taken place with Buckler six years earlier. The gallery expressed its frustration over the continuing limits on its access to the prime minister and his cabinet ministers. MacDougall indicated an openness to discuss improved access, though not to the levels seen before Harper took power, and noted that although the prime minister would continue to avoid interviews around Ottawa, having more media outlets accompany him on international trips was welcomed. In his words, "We'll find ways to make it better and make it work. It's not going to be everything to your liking. It's not going to be everything to my liking. But that's the way it works" (CPPG 2012, 3).

Although much animosity remains between the two sides, relations appear to have dissipated into a more manageable truce, with both politicians and journalists on the Hill becoming accustomed to the working conditions under the Harper government. Whether this new normal constitutes the supplanting of the CPPG and the establishment of a new gatekeeping standard or simply a disruption to the dominant journalist-driven/economic hybrid gatekeeping system is the focus of the next section.

Political Communication in Canada

In assessing whether we are witnessing the erosion of an institution and a concurrent fundamental change in gatekeeping, we need to consider the simultaneous presence of change and stability factors (Bennett 2004). That there continues to be discord between the PMO and the CPPG is not in itself remarkable. As noted earlier, throughout history, the common denominator defining the relationship between the CPPG and Canadian prime ministers has been the issue of who should be responsible for controlling the flows of political communications to the Canadian public and how these communications are framed. Since coming into power in 2006, the Harper governments have maintained a non-stop election readiness, or permanent campaign, that interweaves politics and government (Flanagan 2007, 2012; Chapter 2 in this book). Two key aspects of this practice have created disruptions in the dominant journalist-driven/economic hybrid gatekeeping system in Canada. The first is the adoption of a strategic communication orientation that seeks to manage interactions with the media by minimizing opportunities for journalists to engage in questioning practices that may elicit ill-considered or improvised responses from the prime minister and other members of the cabinet. The second involves communicating foremost through local rather than national media outlets. The objective of both activities is to protect the government's ability to get its message out to Canadians with minimal intercedence from the CPPG. Despite differences in the specific techniques employed, the same claim holds for most previous prime ministers (Levine 1993).

Disruptions to the gatekeeping system have also arisen from the context within which the relations between the CPPG and successive Harper governments have been manifest. It is an environment that combines pervasive public suspicion and wariness with both government and the media, on the one hand (Bastedo et al. 2011; Bastedo, Chu, and Hilderman 2012; Taras and Waddell 2012b), with widespread access to a plethora of information and communication technologies that can be used for the production, diffusion, and consumption of political communications, on the other hand. For many people, the former has served to undermine the classic adage about media as the fourth estate, replacing it with a view that "the media is not an autonomous, objective and innocent entity with a 'god's eye view' of the world" (Ratuva 2003, 177). The widespread distaste for adversarial journalism and politics of personality reflects, according to Alboim (2012, 53), an enduring public desire for, and valuing of, a kind of journalism that adheres more closely to the romanticized notion of journalists as independent watchdogs.

Drawing on Schudson's (1998) study of the historical progress of citizenship practice in the United States, we posit that the Canadian public's apparent lack of concern about the dysfunctional relationship between the PMO and the CPPG is reflective of a reactive, as opposed to proactive, form of civic engagement. This is a form of citizen practice in which the primary obligation is to scan, or monitor, the informational environment so as to be "alerted on a very wide variety of issues" and to mobilize around issues in various ways when individuals deem action to be necessary (Schudson 1998, 310). Seen in this light, it is not the least bit surprising that claims by the CPPG about the negative democratic consequences of the usurpation of its ability to access the prime minister and his cabinet ministers on its own terms failed to resonate, in any substantive sense, with the Canadian public.

Technological factors also aided the Conservatives in circumventing, albeit to a limited extent, the critical lens of the CPPG by combining enhanced use of social media platforms (e.g., Facebook, Twitter, YouTube) with engagement with local and regional outlets to narrowcast to its established support base. In addition, mobilizing a plurality of communication channels and platforms contributed to the success of consecutive Harper governments in severing the possibility of symbiotic relations with the CPPG, and the framing of such actions as a minor Ottawa-based issue of little national import. Although information and communication technologies have facilitated an exponential increase in the dissemination of political information, this does not *ipso facto* render citizens better informed (Dean 2009; Papacharissi 2010). Claims regarding the revolutionary and disintermediating effects of Internet-based communication platforms need to be approached with caution. It is true that people are increasingly making use of online repositories for their news, but this does not equate with a mass eschewing of the mainstream media. Indeed, online mainstream news outlets and television continue to be the principal sources of information for Canadians (CMRC 2011c; 2011d).

The disruptions outlined above, along with many others not discussed in this chapter, have led some observers to posit that we are in the early stages of a transformation in gatekeeping practices in Canada, and elsewhere, that is characterized by a shifting of power "away from old-fashioned networks and newspapers ... towards, on the one hand, smaller news providers (in the case of blogs, toward individuals) and, on the other, to the institutions of government, which have got into the business of providing news more or less directly" (*Economist* 2005). Journalists' work patterns, press-government relations, and communication channels have each undergone notable changes in the past decade. Placing the tensions between the CPPG and the PMO in their broader historical context and examining the instabilities to which they have given rise through the lens of Bennett's (2004) model suggests that the journalistic, organizational, political, and technological factors perpetuating instabilities have yet to shift the existing hybrid gatekeeping system to a radically new point of equilibrium within the Canadian context. Put simply, it is far from clear whether the highly controlled press-government relations characteristic of successive Harper governments and the concomitant tensions between the PMO and the CPPG have indeed propelled the gatekeeping system away from the dominant journalist-driven/economic hybrid.

Therefore, in responding to the question posed in the chapter title, we are left to conclude that although the CPPG's relevance is being continuously challenged, it is hardly a relic of another time. That said, there are several questions for further research whose findings will go a long way toward elucidating the CPPG's gatekeeping role in coming years and the implications for democracy in Canada. These include: How have the changes implemented by successive Harper governments influenced the CPPG's relationship with the other major political parties? And, relatedly, in what ways are the other parties now seeking to control their interactions with the CPPG? Are the changes introduced by Stephen Harper specific to his style of governance, or will his tactics for structuring and controlling the relationship between the PMO and the CPPG remain in place in the post-Harper years? How does the widespread uptake and use of social media platforms alter the role of the CPPG and other similar entities? What impact is the spread of social media platforms having on the quality of political reporting from the CPPG and other similar entities? The most important question of all, however, continues to pertain, as it always has, to the consequences for citizen engagement of the struggle for control of the political news agenda between the CPPG and the government of the day. For this, there are no easy answers.

NOTES

1 Permission to review the email messages was granted on the condition that they be treated as confidential, used solely for review purposes, and not quoted.

8

Setting the Agenda? A Case Study of Newspaper Coverage of the 2006 Canadian Election Campaign

Elisabeth Gidengil

The preceding chapter's profile of the struggle between the Canadian Parliamentary Press Gallery and the Prime Minister's Office illustrates the governing party's frustration with gatekeepers and its desire to set the agenda. This is an aspect of the permanent campaign, for during election campaigns, all political parties compete to control the issue agenda. They strive to emphasize issues on which they enjoy recognized expertise, and to downplay issues that will hurt them. The media play a crucial part in this struggle to control the election agenda (Semetko 1996; Norris et al. 1999). Most voters experience election campaigns only indirectly, by watching the news on television or by reading about the campaign in the press, and so the media have played a critical role in communicating the parties' messages to voters. However, the media do not simply serve as a neutral transmission belt between political parties and voters. In a very literal sense, they mediate the flow of messages, highlighting some and downplaying others as they distill the day's events into newsworthy stories. And in the process, they advantage some parties – and disadvantage others.

Agenda-setting theories imply that the power of the media lies not in telling people *what* to think but in telling them what to think *about* (Cohen 1963, 13). Agenda setting is about determining the salience of issues. According to agenda-setting theories, the amount of attention devoted to particular issues in the news can influence people's issue priorities (McCombs and Shaw 1972; McCombs 1997; see also Chapters 10 and 11 in this book). Agenda setting takes place when extensive media coverage of an issue increases people's perceptions of the issue's importance. The theoretical underpinnings derive from memory-based models of information processing: issues that have received intensive coverage are more readily recalled

(Scheufele and Tewksbury 2007). Agenda setting occurs because of individuals' need for orientation; people turn to the media to learn about politics (Weaver 1977; Wanta 1997, 101; McCombs 2004).

At their simplest, agenda-setting theories assume that the more coverage an issue receives, the more people will perceive that issue to be the most important issue. But agenda setting is not that simple. As this chapter demonstrates, the extent of agenda setting depends not just on how much coverage different issues get but also on the nature of the issues. There are good reasons to believe that agenda-setting effects will be stronger for issues that are normally off the public's radar screen, and weaker for issues like health that are already matters of public concern as a result of people's daily experiences. Zucker (1978) was the first to make this distinction between "unobtrusive" and "obtrusive" issues. An early Canadian study (Winter, Eyal, and Rogers 1982) lent this argument empirical support: agenda-setting effects were much weaker for inflation than for other less obtrusive issues.

Building on this distinction, Soroka (2002) has developed a typology that predicts the potential for agenda setting based on issue attributes. His typology takes account of the obtrusiveness and abstractness (Yagade and Dozier 1990) of an issue, the duration of coverage (Zucker 1978), and the role of dramatic events (MacKuen and Coombs 1981). *Prominent issues* offer the least scope for agenda setting. These are concrete issues that directly affect a large number of people. They are obtrusive in the sense that a lot of people have direct experience with them and therefore have less need to rely on the news media for information or understanding. Because these issues are already salient, there is limited scope for increased coverage to affect their salience. Soroka characterizes both inflation and unemployment as prominent issues. *Sensational issues* offer the greatest scope for agenda setting. They are media driven and they typically attract intense coverage because of their dramatic quality. Sensational issues are concrete and thus easy for people to grasp. Soroka cites AIDS, crime, and the environment as examples. Comparing newspaper coverage and public opinion from 1985 to 1995, he found some support for his typology. As predicted, inflation and unemployment proved to be real-world led, whereas the environment was clearly media driven. However, despite their sensational nature, crime and AIDS revealed few agenda-setting effects.

Agenda setting is potentially consequential because it can affect voters' decision calculus. This is what is known as priming. Priming occurs when media attention to an issue causes people to place special weight on it when evaluating candidates or deciding which party to support. We can think of priming as "the electoral manifestation of the elite struggle for control of the agenda" (Johnston et al. 1992, 212; see also Iyengar and Kinder 1987). Priming is a form of persuasion. It can lead people to change their minds, not because they have changed their opinions but because the relative weight of those opinions in their decision has changed. This is precisely what motivates the parties' struggle for control of the agenda: "Politicians prime issues to provide people with reasons for supporting them" (Johnston et al. 1992, 5).

Political parties typically want to prime the issues that they "own." According to the theory of issue ownership (Budge and Farlie 1983; Petrocik 1996; Bélanger 2003), some parties are viewed as more competent than others when it comes to dealing with a particular issue. For example, voters tend to view parties of the left as best for dealing with welfare, education, and unemployment, whereas parties of the right are often perceived as strongest on national defence, crime, and inflation (Budge and Farlie 1983; Petrocik 1996; Kaufmann 2004). These images of party competence – and incompetence – on particular issues tend to be relatively enduring, and so political parties have strong incentives to play up issues that they "own" and to play down those "owned" by their rivals.

Whether issue ownership matters to vote choice depends on the issue's perceived salience (Bélanger and Meguid 2008). Issue ownership affects only the voting decisions of voters who consider the issue important. There is no reason to expect that thinking that the Conservatives, say, would be the best party for dealing with crime will affect people's vote unless they are concerned about crime. The greater the perceived salience of the issue, the more impact issue ownership has on people's choice of party. Thus, media coverage can indirectly influence people's vote by shaping their perceptions of an issue's importance.

The media's agenda may well differ from the parties' agendas. There is only a finite amount of space in the newspaper or time on the air, and so agenda setting is apt to be a zero-sum game: the more coverage one issue receives, the less coverage there will be of other issues. André Blais and his colleagues (2002) have compared the attention devoted to various issues in parties' press releases with the amount of attention those issues received in the English-language network news during the 2000 federal election campaign. Coverage of the Liberals corresponded quite well to the issues that the party was highlighting in its press releases, namely health and public finance. However, news coverage paid very little attention to the Liberals' economic agenda, and a lot of attention to coverage of ethical issues. The latter was precipitated by the prime minister's acknowledgment that he had talked to the chair of the Business Development Bank of Canada on behalf of an investor who was seeking a loan for a hotel in the prime minister's riding. The Canadian Alliance's greatest media success was focusing media attention on ethical practices and away from the constitutional question, which had hurt Reform, the party's predecessor, in 1997. On the other hand, the party failed to achieve its goal of putting the spotlight on public finances. Worse, the focus on health put the Alliance on the defensive over its supposed support for two-tier health care. Meanwhile, to the extent that it got covered, the New Democratic Party's coverage highlighted the issues that dominated its press releases – public finance, social programs, and above all, health.

Case Study

Does the amount of media coverage afforded different issues influence voters' issue priorities? This question has attracted little attention in the Canadian context. This case study begins to fill this gap by examining the relationship between the intensity of newspaper coverage of various election issues and the perceived importance of those issues to voters. In focusing on the press, the case study is, in effect, looking for agenda-setting effects where we might most expect to find them: "Newspapers are the traditional agenda-setting medium, and they have also been the barometer par excellence of mass media's message and influence for political scientists" (Andrew 2007, 28). The setting is the 2006 federal election campaign in English-speaking Canada. Election campaigns are critical moments for analyzing agenda setting. This is when the stakes are highest and the competition between the parties to set the agenda is fiercest. It is also when people's need for orientation from the media is greatest.

The 2006 election was called on November 29, 2005, following the defeat of the minority Liberal government on a non-confidence vote. To allow for the Christmas and New Year holidays, the campaign lasted almost eight weeks. The two major contenders adopted very different strategies. Whereas the Liberals did not make any major policy announcements until after the holidays, the Conservatives announced a policy per day well into January. The Conservative strategy meant that issue coverage during the first half of the campaign focused heavily on these pronouncements (Waddell and

Dornan 2006). Nonetheless, it was the Liberals who emerged as the net winners on the issues (Gidengil et al. 2012), though this was not sufficient to keep the party in power, and the election resulted in a Conservative minority government.

METHOD

A content analysis of daily press coverage is combined with rolling cross-section survey data to determine whether there is a systematic relationship between changes in the intensity of newspaper coverage of various issues and the priority that Canadians attached to those issues. A rolling cross-section design breaks the total sample down into daily samples (see Johnston and Brady 2002). Because each daily sample is as similar to the others as random sampling permits, all that distinguishes them (within the range of sampling error) is the date of interview. Combining daily tracking of newspaper coverage with daily sampling of eligible voters provides a powerful design for capturing the dynamic interaction between media coverage and voters' reactions (Johnston, Hagen, and Jamieson 2004).

This design is particularly well suited to studying agenda setting because it can establish whether changes in media coverage are followed by changes in voters' priorities. This temporal sequence is essential for inferring causation: high correlations between the media agenda and voters' issue priorities could simply mean that the media were able to achieve a good fit between their messages and the interests of their audience (McCombs and Shaw 1972, 184-85). Daily tracking also means that confounding events pose much less of a threat than is the case when there is a lengthy elapse between the waves of a longitudinal survey. Finally, unlike controlled experiments, the rolling cross-section design allows us to observe the ebb and flow of citizens' priorities in a natural setting with real issues.

The survey data are taken from the 2006 Canadian Election Study (CES).[1] Interviewing started the day the election was called on November 29, 2005, and (except for December 24, December 25, December 31, and January 1) continued until the eve of the election, which was held on January 23, 2006. At the beginning of the survey, respondents were asked "What is the most important issue to you personally in this federal election?" Separate dependent variables were created, one for each issue, coded "1" for mentions of the issue and "0" for all other mentions and don't knows combined.

The analysis excludes Quebec respondents. Differences in the party system and in the way that election campaigns unfold in Quebec are reflected in voters' issue priorities: for example, according to the 2006 CES, only 26 percent of Quebeckers named health as the most important issue, and a mere .1 percent named crime, compared with 33 percent and 4 percent respectively of people living outside the province.

The data on press coverage are derived from a content analysis of election stories in the five major Canadian daily newspapers: the *Globe and Mail, National Post, Toronto Star, Vancouver Sun,* and *Calgary Herald.* The content analysis was automated using Lexicoder.[2] This software tool codes media content based on a dictionary of keywords for seventeen major policy domains. A story that mentions at least two topic keywords is assumed to contain coverage of the given domain. Every story (including editorials and opinion pieces) in the main news section of each newspaper that mentioned the federal election or any of the parties or party leaders was coded. The analysis is based on pooled data for all five newspapers.

Pooling implicitly assumes that there is a Canadian newspaper agenda. Given that media coverage responds to real-world events, this seems a reasonable assumption, particularly in the campaign context where press releases and party leaders' pronouncements help drive coverage. The evidence certainly points in this direction (see, for example, Frizzell and Westell 1989; Soroka 2002), but it needs to be established empirically, especially given that the sample of newspapers does not span every part of the country. Accordingly, two tests have been performed. The first involves calculating Cronbach's alpha. This measure was developed to test the reliability of attitudinal scales, but it can also serve to summarize the extent to which the salience of various issues in different newspapers increases and decreases in tandem (see Soroka 2002).

Whether it is possible to speak of a common issue agenda depends very much on the issue at hand (c.f. Soroka 2002). Inter-newspaper similarity is greatest for foreign affairs and defence, civil rights, ethics, crime and justice, finance and commerce, taxes, and health (see Table 8.1). All have coefficient alphas of .60 or higher, and deleting one of the newspapers generally does not result in a higher value. Turning to issues with lower coefficients, it is difficult to discern any consistent patterns relating to ownership or region.[3] Instead, lack of consistency across newspapers likely reflects local particularities. It is hardly surprising, for example, that a Toronto-based newspaper is an outlier on coverage of agriculture. The criterion for extraction was an eigenvalue of 1.00 or higher. The second test yields similar conclusions.

TABLE 8.1

Consistency in issue coverage in five Canadian newspapers

Issue	Alpha	Alpha if item deleted	
Agriculture/forestry/fishing	0.48	0.58	*(Toronto Star)*
Civil rights	0.72	0.71	*(Calgary Herald/Toronto Star)*
Crime/justice	0.69	0.68	*(Toronto Star)*
Economy	0.54	0.53	*(Toronto Star)*
Education	0.40	0.40	*(Calgary Herald)*
Employment/labour	0.32	0.32	*(Toronto Star)*
Energy	0.42	0.56	*(Vancouver Sun)*
Environment	0.46	0.54	*(Globe and Mail)*
Ethics	0.70	0.72	*(Globe and Mail)*
Finance/commerce	0.69	0.74	*(Globe and Mail)*
Foreign affairs/defence	0.73	0.73	*(Globe and Mail)*
Health	0.64	0.62	*(Calgary Herald)*
Immigration	0.57	0.62	*(Toronto Star)*
Social welfare	0.25	0.33	*(Vancouver Sun)*
Taxes	0.66	0.68	*(Vancouver Sun)*
Trade (international)	0.54	0.54	*(Vancouver Sun)*
Transportation	0.03	0.25	*(Vancouver Sun)*

NOTE: Cells contain Cronbach's alphas, using daily data.

Table 8.2 presents the results of a principal components analysis (see Soroka 2002). Principal components analysis involves identifying patterns of intercorrelation among a set of items. Components are extracted based on how much of the variance they explain, beginning with the component that explains the most. If the amount of coverage accorded a given issue follows a similar trend, the newspapers will all load heavily on a single component. This is the case for civil rights, ethics, foreign affairs and defence, health, and taxes. On the other hand, the amount of coverage accorded to the economy, the environment, social welfare, and trade clearly follows different dynamics, depending on the newspaper. The result for crime and justice is more ambiguous: the *Toronto Star* has a relatively high loading on the first component, yet its highest loading is on the second component. Overall, though, there is sufficient inter-newspaper similarity to warrant pooling the data, at least for a subset of issues.

But did voters become more concerned with these issues when they received greater coverage? For media coverage to have an effect on people's

Table 8.2

Regional and ownership trends in press coverage

Newspaper		Calgary Herald	Globe and Mail	National Post	Toronto Star	Vancouver Sun	% variance
Agriculture, etc.	1	0.670	0.595	0.738	−0.219	0.694	37.5
	2	0.354	−0.345	0.379	0.803	−0.196	21.4
Civil rights	1	0.623	0.801	0.786	0.601	0.642	48.4
Crime/Justice	1	0.782	0.727	0.686	0.574	0.583	45.6
	2	−0.451	0.137	−0.518	0.602	0.451	21.1
Economy	1	0.730	0.631	0.509	0.395	0.656	35.5
	2	0.243	−0.582	0.792	0.185	−0.437	25.0
	3	−0.526	0.024	−0.002	0.870	0.040	20.7
Education	1	0.391	0.433	0.580	0.665	0.630	30.3
	2	0.593	0.592	0.093	−0.488	−0.344	21.3
Employment	1	0.533	0.449	0.708	0.370	0.509	27.6
	2	0.380	−0.699	0.121	0.645	−0.419	24.8
Energy	1	0.591	0.830	0.597	0.727	0.769	35.6
	2	0.433	0.321	−0.230	−0.367	0.826	23.2
Ethics	1	0.800	0.465	0.735	0.830	0.568	48.2
Environment	1	0.722	0.000	0.605	0.643	0.621	33.7
	2	−0.344	0.947	0.173	0.141	0.085	21.4
	3	0.330	0.168	−0.498	0.604	−0.523	20.5
Finance	1	0.822	0.296	0.675	0.726	0.747	46.1
	2	−0.074	0.901	−0.402	0.222	−0.128	20.9
Foreign	1	0.769	0.591	0.830	0.597	0.727	50.3
Health	1	0.542	0.667	0.675	0.669	0.701	42.6
Immigration	1	0.564	0.614	0.729	0.154	0.808	38.1
	2	−0.455	0.444	−0.009	0.859	−0.176	23.5
Taxes	1	0.720	0.668	0.756	0.651	0.494	44.1
Trade	1	0.835	0.507	0.520	0.383	0.664	36.3
	2	−0.052	0.644	−0.357	0.642	−0.519	24.5
Transportation	1	0.496	0.782	−0.346	0.641	0.135	28.2
	2	0.600	0.088	0.795	−0.040	−0.483	24.7
	3	0.466	−0.227	0.167	−0.167	0.826	20.2
Social welfare	1	0.387	0.659	0.837	0.313	−0.102	27.9
	2	0.437	−0.570	−0.060	0.673	−0.451	23.5
	3	0.606	−0.035	−0.192	0.099	0.805	21.3

NOTES: Cells contain factor loadings from an unrotated principal components analysis, using daily data. Figures in semibold indicate the highest factor loading for each newspaper on the given issue.

perceptions of an issue's importance there must be variation across time in the amount of coverage that the issue receives, and there must be periods of intense coverage. If an issue receives very little coverage, it is unlikely to register with most of the public unless and until something causes a spike in media attention. Analogously, if the issue receives sustained coverage over a lengthy period, the public's perception of the issue's importance is unlikely to change.

Findings

Figure 8.1 compares trends in coverage of the issues for which there is a common newspaper agenda with trends in respondents' identification of the most important issue. The finance/commerce issue is excluded because few, if any, respondents named it as their most important issue. Given the small daily sample sizes, five-day moving averages are used to smooth the tracking. The graphical data provide some initial support for the agenda-setting hypothesis, at least in the second half of the campaign. Following the holiday break, newspaper coverage of issues relating to civil rights steadily increased and so did the proportion of people saying that rights-related issues were the most important to them personally. This was not a case of press coverage picking up on what mattered to voters. On the contrary, the increase in perceived importance clearly followed the increase in coverage. The pattern is similar for issues relating to crime and justice, at least in the second half of the campaign: increased press coverage appears to be driving growing public concern. Note that both issues qualify as "sensational" (Soroka 2002) and, in the case of crime/justice, it was clearly a dramatic event – the Boxing Day murder of a Toronto schoolgirl – that sparked a flurry of newspaper stories. Ethics is also a sensational issue, and there is some evidence – albeit modest – of an agenda-setting effect in the second half of the campaign. This was very likely related to another dramatic event: the announcement on December 28 that the RCMP were launching a criminal investigation into whether details of a decision on taxing income trusts had been leaked from the finance minister's office prior to the announcement, as alleged by the NDP's finance critic.

Clearly, though, there is no necessary connection between the intensity of newspaper coverage and the perceived importance of issues. Intense coverage of civil rights and crime/justice in the first half of the campaign is not reflected in a corresponding shift in the perceived importance of these issues. Conversely, a peak in public concern for ethics in mid-December is unrelated to any shift in the intensity of coverage.

Figure 8.1

Trends in newspaper coverage and perceived importance

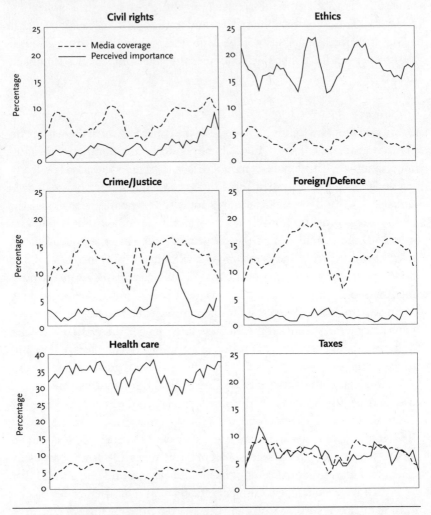

The results for health confirm the obtrusiveness hypothesis: issues that are already on people's minds typically offer the least scope for media agenda setting (Zucker 1978). Health is the quintessential prominent issue, and so it is unsurprising that it was by far the most frequently mentioned issue despite receiving relatively little coverage throughout the campaign. Note, though, that a dramatic event, like media reports in 2000 that the Alliance would introduce two-tier health care, can enhance the salience of a prominent issue. It is also unsurprising that shifts in foreign affairs/defence coverage do not appear to have registered with the public. During an election campaign, voters' minds are focused on domestic issues, and foreign affairs can seem abstract in the absence of some sensational event like a terrorist attack. The pattern for taxes is more surprising: despite the Conservatives' attempts to make it an issue, taxes did not receive much attention in the press *or* elicit much concern on the part of most voters.

Simply comparing trends in coverage and public concern can be misleading. Regression analysis offers a way of establishing that agenda-setting effects hold while controlling for the effects of other factors that could plausibly affect the dynamics of perceived issue importance. The independent variables correspond to periods of intense coverage.[4] This reflects a threshold approach where "a minimum intensity of reporting is necessary to exert an effect on perceived issue importance" (Brosius and Kepplinger 1992, 8). It may be unrealistic to assume that intense media coverage has an immediate effect and that the effect vanishes as soon as coverage diminishes. It is more plausible to think of media coverage as having a cumulative effect (Fan 1988): it may take a few successive days of intense coverage to influence voters' perceptions of the relative importance of an issue. Support for this notion comes from Iyengar and Kinder's (1987) experiments, which showed that periodic exposure spread over several days had more impact than a single concentrated exposure. Accordingly, the media variables accommodate some lag-time for the effects to register and some time for the effects to decay. Specifically, the models assume that the effects of intense media coverage will grow in .33 daily increments after each issue becomes salient, until they reach the value of 1.0 on the third day of intense coverage, and then decline in .33 daily increments to zero after the end of the period when the issue was salient.

The models also control for other possible sources of perceived issue importance. For example, women are typically more concerned about health care and less concerned about taxes than are men (Gidengil et al. 2003); Christian fundamentalists are more likely to think that matters such as abortion and same-sex marriage are important; Conservatives are more likely to care about taxes; the politically disaffected are more likely to be concerned with ethics, and so on. The social background characteristics were all dummy-coded with the named category coded 1. In the case of party identification, one dummy variable was created for each party, with weak

TABLE 8.3

The effect of intense press coverage on perceived issue importance

	Civil rights	Crime/justice	Ethics	Foreign affairs/ defence
Intense coverage$_{t1}$	−1.38 (.53)**	0.57 (.81)	—	0.00 (.62)
Intense coverage$_{t2}$	0.76 (.32)*	0.94 (.33)**	0.26 (.14)^{a}	−0.17 (.73)
Under thirty-five years of age	−0.45 (.44)	0.02 (.36)	−1.12 (.18)***	0.08 (.45)
Over fifty-four years of age	0.17 (.32)	0.19 (.39)	0.32 (.13)*	−0.05 (.43)
Atlantic resident	−0.96 (.48)*	−1.58 (.64)*	0.11 (.20)	−0.01 (.55)
Western resident	−0.05 (.23)	−0.48 (.22)*	0.52 (.12)***	−0.58 (.36)
Female	0.39 (.28)	−0.25 (.19)	−0.75 (.13)***	−0.30 (.42)
University graduate	−0.16 (.32)	0.20 (.27)	0.32 (.13)*	0.38 (.33)
Public-sector employee	0.49 (.46)	0.61 (.42)	0.11 (.18)	0.23 (.42)
Union household	−0.39 (.32)	−0.28 (.40)	−0.05 (.16)	−0.42 (.40)
Conservative Party identification	0.93 (.31)**	−0.04 (.23)	0.65 (.14)***	0.02 (.62)
Liberal Party identification	0.56 (.39)	−0.44 (.28)	−0.31 (.17)^{a}	0.43 (.52)
NDP identification	0.43 (.40)	−1.82 (.64)**	−0.40 (.22)^{a}	0.49 (.62)
First debate	1.40 (.42)***	−0.99 (.76)	0.68 (.26)*	−0.07 (1.36)
Second debate	−0.50 (.83)	0.77 (.40)^{a}	−0.09 (.18)	−2.21 (1.04)*
Christian fundamentalist	1.07 (.28)***	—	—	—
Political disaffection	—	—	0.05 (.03)*	—
Constant	−4.69 (.38)***	−3.35 (.33)***	−1.72 (.16)***	−3.99 (.38)***
Nagelkerke's pseudo R-squared	0.10	0.09	0.13	0.03
Number of cases	2,693	2,794	2,775	2,794

NOTES:

Column entries are unstandardized logistic regression coefficients, with robust standard errors in parentheses.

t1 and t2: The "t" refers to periods of intense coverage. For some issues, there were two periods. The length and timing of these periods vary depending on the issue.

$^{a}p <.10$; $^{*}p < .05$; $^{**}p < .01$; $^{***}p < .001$

identifiers and non-identifiers serving as the reference category (see Blais et al. 2001). The jury is still out on how much televised leaders' debates matter (see Gidengil 2008), and studies of debate effects have not looked at their impact on agenda setting. However, it is reasonable to assume that debates could influence voters' perceptions of issue importance, if only temporarily. Accordingly, the analysis also takes account of the two English-language televised leaders' debates that took place on December 16 and January 9.

Since it can take a few days for the full effect to register with voters (Blais and Boyer 1996), the debate variables take the value of zero until the day of the debate and then progressively increase to 1 in the following three days before declining back to zero over the next three days.

Even taking account of other factors that may affect their perceived importance, there is clear evidence of agenda setting for civil rights, crime/justice, and ethics (see Table 8.3). As the graphical representations indicated, the agenda-setting effects are confined to the second half of the campaign. The effects are strongest for crime/justice and ethics. The estimated probability of identifying crime/justice as the most important issue is 3.5 points higher during the period of intense coverage after the holidays, as is the probability of naming ethics. The effect is weaker for rights-related issues: the estimated probability is 2.1 points higher during the period of intense coverage in the latter part of the campaign. At the same time, the regression models confirm the lack of any agenda-setting effect for news about foreign affairs/defence. The two periods of intense press coverage failed to register with the electorate.

The regression models also help explain why public concern for ethics peaked in mid-December, despite the lack of coverage. The spike in public concern was clearly associated with the first televised leaders' debate on December 16. The debate also served to enhance the perceived importance of rights-based issues. There was increased press coverage around the same time (see Figure 8.1), but it did not have any effect over and above the debate.

These findings raise an interesting question: Why are the agenda-setting effects confined to the second half of the campaign? The most likely answer has to do with the amount of attention that people were paying to the campaign. Indeed, we might expect that agenda-setting effects at the individual level would depend on how much attention people are paying to the news. This turns out, though, not to be the case: the effects observed here hold regardless of how much attention people were paying to news about the upcoming election.[5] This seemingly counterintuitive finding can be explained in terms of the classic two-step flow of communication hypothesis (Lazarsfeld, Berelson, and Gaudet 1948). This hypothesis recognizes that media effects are both direct and indirect. Direct effects result from actual exposure to media coverage, whereas indirect effects are produced as people discuss with their friends, acquaintances, and family members what they have read or heard (see, for example, Blais and Boyer 1996). In other words, the agenda-setting power of the media may well have been amplified as people talked about the election at the water cooler, over the back fence, or around the dinner table (for evidence of this in the context of priming, see Gidengil et al. 2002).

As Figure 8.1 indicates, there were no periods of intense coverage for health care and taxes. The disjuncture between the importance of health care to many voters and the modest amount of press coverage is striking. Health care is a prime example of a prominent issue. It was on many voters' minds from the get-go, with women, Liberals, and New Democrats being especially likely to think that health care was the most important issue. In the case of taxes, press coverage and public concern alike fluctuated across the campaign in no discernible pattern. To the extent that it was a salient issue in voters' minds, it was very much the preoccupation of men and Conservatives.

Political Communication in Canada

When a given issue received intense press coverage, there was a modest but significant increase in the number of people who spontaneously named that issue as the most important to them personally in the election. However, the effects were confined to the sensational sorts of issues that lend themselves to agenda setting on the part of the media. Clearly, there are very real limits to the press's ability to tell people what to think about. Indeed, the case study may be overstating the extent of agenda setting in Canadian elections; studies have shown that, half the time, newspapers and television had the same agenda-setting effects, but in the remaining cases, newspapers had stronger agenda-setting effects by a ratio of about two to one (McCombs 2004, 49).

The limited scope of agenda setting by the press may help explain why issue priming seems to be the exception rather than the norm in Canada, occurring only when a new and dramatic issue dominates the campaign, as it did in the 1988 election, which revolved to an unusual degree around a single issue: free trade with the United States (Johnston et al. 1992; Gidengil et al. 2002). It also helps explain why the issues of the day have only a limited effect on voters' choice of party (Gidengil et al. 2012).

Why are the agenda-setting effects observed here not stronger? The 2006 campaign was longer than usual, but it may still be that the official election period is simply too short to allow much scope for agenda setting on the part of the press (Norris et al. 1999). Studies of agenda setting have typically assumed that it takes several weeks of coverage for effects to show up

(McCombs 2004, 44). Moreover, election coverage in the print and broad-cast media typically does not pay much attention to the issues (Mendelsohn 1993; Waddell and Dornan 2006). The focus is on campaign gaffes and the horse race, who is ahead in the polls and who is behind, and when issues are covered, it is often in a way that is derivative of the horse race or strategic frame: "Policy positions are treated as mere campaign devices to attract votes" (Mendelsohn 1993, 158).

Another possible reason agenda-setting effects are not stronger is the regionalized nature of Canadian election campaigns. This would help explain the lack of a Canada-wide agenda on several issues in the English-language press. This lack also probably reflects regional differences in voters' priorities (see Gidengil et al. 1999). Market constraints mean that newspapers are go-ing to devote coverage to the issues that concern their readers, and some of these are going to be local in nature. For press coverage to affect the issue priorities of voters at large, an issue has to gain attention from newspapers across the country. During the 2006 election campaign, for example, the Boxing Day shooting death of a schoolgirl in Toronto was sufficiently sensa-tional that it figured in nation-wide press coverage, and this intense coverage was reflected in voters' heightened concern with crime.

Intense coverage is necessary for agenda setting to occur, but it may not be sufficient. Many other factors besides the amount of media coverage can affect the public's perceptions of the most important issues, including their social backgrounds and partisan predispositions. Voters are not blank slates; they have predispositions that may well influence the salience of particular issues quite independently of the amount of coverage those issues receive. This is especially true of prominent issues such as health care.

A final possible reason for the weakness of agenda setting may be the proliferation of news sources: the greater the diversity of news sources, the less scope there will be for agenda setting by the media. Indeed, the prolif-eration of online sources of news and information has led some observers to suggest that the print and broadcast media will cease to play a role in setting the agenda (Takeshita 2006). Political parties no longer have to rely so heav-ily on the traditional media to get their messages out to voters. Email, web-sites, social networking sites like Facebook, and microblogging services like Twitter allow parties and candidates to communicate directly with voters and to tailor their messages to particular audiences. For their part, voters have unprecedented access to news and information about politics on the

Internet and can personalize their news content to focus on those issues they care about.

Some observers argue, though, that claims about the demise of agenda setting have been greatly exaggerated (McCombs 2005). Large numbers of people still rely on the traditional broadcast and print media for their news. This is even true of those who report reading the news online every day. According to the Canadian Election Study (2011), 54 percent of those who read the news online every day also watch the news on television every day; 46 percent listen to the news on the radio every day; and, surprisingly, 39 percent read the news in the newspaper daily. Moreover, when people go online for their news, they will often be visiting the websites of major newspapers, news magazines, and television networks. Real-world events, along with journalistic norms and economic imperatives, may well mean that political blogs and other online news sites are also presenting fairly similar content (McCombs 2005; Lee 2007). Where online sources of news and information may really come into play is in inter-media agenda setting. It is possible, of course, that the agenda-setting role of the traditional print and broadcast media will diminish as more alternative news sources become available and as more people turn to online sources for their news about politics.

The observed agenda-setting effects may be limited, but most studies have assumed that these effects occur only after several weeks of media exposure (McCombs 2004, 44). The evidence from the 2006 Canadian election suggests otherwise: press coverage can influence voters' issue priorities even within the short time span of an election campaign. But conditions have to be right. Agenda setting is most likely to occur when a dramatic event captures the attention of newspapers across the country.

NOTES

1 The survey was conducted by the Institute for Social Research at York University using computer-assisted telephone-interviewing software. The average interview was thirty minutes. A total of 3,045 Canadian citizens were interviewed outside Quebec. The response rate was 61 percent. Of the respondents, 1,567 of them had also been interviewed as part of the 2004 Canadian Election Study. To account for any possible differences in issue priorities, the analyses include a control for 2004-6 panel respondents. The data have been weighted to correct for differences in the probability of selection based on household size and province of residence. Copies of the questionnaire, technical documentation, and data are available at www.queensu.ca/cora/ces.html.

2 Lexicoder was developed at the McGill Institute for the Study of Canada by Lori Young and Stuart Soroka, and programmed by Mark Daku. I am grateful to Stuart Soroka for providing the data. He does not bear any responsibility for the interpretations presented here. The Lexicoder Topic Dictionary is available at www.lexicoder.com (see also Soroka et al. 2011; Andrew, Fournier, and Soroka 2013). The topic of government/politics is not included in the analysis, since it does not qualify as an issue. The topic of ethics was added to the dictionary, using the following keywords: "corruption," "income trust," "ethics," "Gomery," "adscam," and "sponsorship scandal."

3 At the time, CanWest Global owned the *Calgary Herald, National Post,* and *Vancouver Sun.* Bell Globemedia owned the *Globe and Mail,* and Torstar owned the *Toronto Star.*

4 More specifically, December 9 was the first day crime figured prominently in more than 15 percent of newspaper stories about the election; December 12 was the last. A second period of intense coverage followed, beginning on December 28 and ending on January 15. For civil rights, there were also two periods of consistently high coverage days: from December 17 to December 19, and again from January 9 to January 22, this topic accounted for more than 10 percent of news stories. The high coverage days for ethics lasted from December 30 to January 10. However, this topic never featured on a daily basis in more than 9 percent of stories about the election. For foreign affairs/defence, there were two periods of intense coverage: from December 10 to December 22, and again from January 13 to January 16, when the issue accounted for more than 15 percent of news stories.

5 Following Zaller (1992), a measure of campaign-relevant knowledge was used to test for the effects of media attention. Whether people could correctly identify the leaders of all three federal parties made no significant difference to the impact of surges in press coverage of the various issues.

9

Playing along New Rules: Personalized Politics in a 24/7 Mediated World

Mireille Lalancette, with Alex Drouin and
Catherine Lemarier-Saulnier

Since the late 1990s, the personalization of politics phenomenon has been the subject of a growing interest from researchers all over the world.[1] As Van Santen and Van Zoonen (2009) and Van Aelst, Sheafer, and Stanyer (2011) illustrate, researchers offer many conceptualizations of the phenomenon. For example, individualized personalization is characterized by an increased attention from the media to the private life of politicians, whereas institutional personalization concerns situations where individual representatives tend to embody the institution itself. The personalization of politics phenomenon could therefore be considered as a normal process where personalities embody collective battles. For instance, the past Quebec-Canada political struggles have been presented as a fight between Quebec premier René Lévesque and Prime Minister Pierre Trudeau (A.-M. Gingras 2009). However, another hypothesis is that personalization has increased since the 1970s, following the election of Margaret Thatcher and Ronald Reagan, in which cases strong, charismatic leaders have eclipsed their respective parties (McAllister 2007). Yet, for McAllister, personalization can actually be traced to Trudeau's election in 1968 and Trudeaumania.

Combining several existing definitions, we understand personalization as the emphasis placed on political actors, their personal qualities, and their private lives behind the scene (place of birth, family background, previous occupations, relationship with spouse and children, personal tastes, and hobbies). The phenomenon of personalization of politics is often associated with media malaise literature and with concepts such as tabloidization and infotainment (for example in Blumler and Kavanagh 1999; Mazzoleni and Schulz 1999; Norris 2000; Neveu 2005). This association results in a negative assessment of political news coverage in which personalization is accused of drawing attention away from policies, political programs, and parties in favour of the personal features of politicians (Van Santen and Van Zoonen 2009).

As McAllister (2007) notes, the formal and informal rules that govern politics tend to change with personalization, along with the transformation of the media environment and of political communication practices. First, the most popular explanation points to the ongoing expansion of the media environment since the 1950s and 1960s. More specifically, television brought politicians inside citizens' home, giving them a face and a voice, and thus promoting the personalization of politics (Campus 2010). Second, in a 24/7 environment of constant news production, politicians need to make greater efforts to attract the attention of both media and the public (Holtz-Bacha 2002; Corner and Pels 2003; Abélès 2007; Goodyear-Grant 2009; McNair 2011). The growing role of political communication professionals also influences this trend (Negrine and Lilleker 2002). Third, some hypotheses emerge about the changing nature of politicians. When being "mediagenic" becomes mandatory, party leaders are more likely to be selected based on their ability to deal with the media than on their skills at knitting alliances among political factions and social groups (Brants and Voltmer 2011, 6). Image-control strategies and the privatization of politics (i.e., the presentation of candidates in their private environment rather than in their political role) are becoming common practice (Holtz-Bacha 2002; Langer 2012; see also Chapter 4 in this book). In this sense, personalization is also tied to the celebratization (see, for example, Hart and Tindall 2009; Trimble and Everitt 2010) and the intimization of politics (Stanyer 2013). Fourth, "political representation is an art that draws on the skills and resources which define mass-mediated political culture" (Street 2004, 446). In such a context, politicians become characters in media dramas, in which narratives derive from "myth and popular heroes, sliding between life and art" (ibid.; Dakhlia 2009, 2010). Fifth, politicians must devote substantial efforts to the construction of their political persona (Van Zoonen 2006), trying to present themselves through every performance on the public scene as a person of quality (Marshall 1997; Corner 2003). This is especially relevant during leadership races, where contenders aspire to be the leader of the party, as we illustrate later in this case study. Finally, several other factors do make a difference in the personalization of politics. For example, the very nature of a political system influences the way politicians perform, as presidential and parliamentary systems operate differently. Similarly, media organizations and their operating rules often vary from one country to another.

In reaction to the trend being often taken for granted, political communication researchers have tried to empirically understand and document issues of personalization.[2]

They have done so in several democratic countries: Kriesi (2012) in Austria, France, Germany, the Netherlands, Switzerland, and the United Kingdom; Langer (2010) in Britain; Morris and Clawson (2005) in the United States; Rahat and Sheafer (2007) in Israel; Reinemann and Wilke (2007) in Germany; and Vliegenthart, Boomgaarden, and Boumans (2011) in the United Kingdom and the Netherlands. However, all these scholars stress the importance of carrying out more research to understand the forms, formats, and qualities of the personalization phenomenon. This is therefore one of the more specific objectives of our case study.

As usually staged and spectacularized events, leadership races appear as a fertile ground to compare media coverage of candidates (Trimble 2007). Unlike general elections where parties compete, leadership races are focused on leaders and are therefore a great way to document the nature of personalization processes in the press. As top candidates get the media talking, their past, present, and future actions; their personalities; their private lives; and their families are scrutinized and evaluated by the party, their members, their opponents, the citizens, and the press. Therefore, what are the most important forms (political and private) of personalization coverage during leadership races? Has media personalization of political leaders increased in Canada over the years?

This case study also documents the personalization process related to changes in criteria used to assess politicians. The existing literature on gender, media, and politics is primarily about the relation between gender, media, and politics. Researchers from that field do not mobilize the key literature on personalization, nor do they assess whether the phenomenon is increasing in the news. Rather, they are centred on gender bias and the difference of coverage between male and female politicians, and on the possible effects of it on political participation and female enrolment in politics. So even if this literature is not studying personalization per se, its results are useful to study how professional competences and performances, as well as personality traits, are presented and mobilized to assess candidates. Their focus on the way media present male and female politicians differently, and emphasize their backgrounds and private lives, resonates with our case study. For example, women are frequently asked how they will balance marriage, motherhood, and politics, which is still considered a non-conventional career choice for women (Norris 1997b; Van Zoonen 1998, 2006). Also, in the new Conservative Party leadership race of 2004, readers were constantly reminded that Belinda Stronach was first and foremost a woman, while considerable attention was drawn on the marital status and family life of the candidates (Trimble 2007; also confirmed by Heldman, Carroll, and Olson 2005; Lalancette and Lemarier-Saulnier 2013). The focus on the novelty status of female candidates (Van Acker 2003; Trimble, Treiberg, and Girard 2010) and on their personality also revealed gender bias (Lawrence and Rose 2010), as well as many double binds for women (K.H. Jamieson 1995; Lemarier-Saulnier and Lalancette 2012; Lalancette and Lemarier-Saulnier 2013).

Since leadership is still assumed to be masculine territory (Trimble and Arscott 2008), academics have also focused on leadership races to identify gendered patterns. Bashevkin (2009b) studied the effects of the party's influence, as well as the individual effects of female candidates, in federal party leadership races between 1975 and 2006. Everitt and Camp (2009a, 2009b) examined the media coverage of the 2005 New Brunswick NDP leadership race. Their focus was on the framing of Allison Brewer's candidacy, which led them to identify the many labels used to describe this "out" lesbian aiming for public office. Trimble (2007) looked at the way the *Globe and Mail* covered three Conservative Party of Canada leadership races featuring female contenders (1976, 1993, and 2004). In a related field, O'Neil and Stewart (2009) compared the experiences of male and female party leaders at the provincial and federal levels in Canada between 1980 and 2005. They found that men enjoy greater electoral success and longer tenures. In the United States, Heldman, Carroll, and Olson (2005) studied the coverage of Elizabeth Dole's attempt to win the Republican presidential nomination and how the coverage was not gender-neutral.

This brings us to additional questions, such as whether a gender component exists and, if so, to what extent the introduction of a female candidate changes the way the race is covered – and finally, what changed and what remained stable in the personalization processes in recent years. To answer these questions, we explore the newspaper coverage of a selection of political party leadership races in Quebec and Canada from 1989 to 2008.

Case Study

Method

For this case study, we conducted a qualitative and quantitative content and discourse analysis of news stories published in French-language newspapers and public affairs magazines in Quebec on nine political party leadership races (see Table 9.1). The publications selected were *L'Actualité, La Presse, Le Devoir, Le Soleil, Le Nouvelliste*, and *Voir*. They were chosen because they are the major publications in Quebec and in the region of our study. These races were selected to portray most parties in Quebec and Canada, and to facilitate comparisons, as several include both male and female contenders. To assess the importance of gender issues, we selected races where female candidates won, as well as ones where only male contenders were in the top three candidates. Also, we chose leading, Opposition, and third parties to ensure a variety of viewpoints and potential processes of personalization in the press.

In this research, we chose to focus only on media published in Quebec in order to ensure a cultural coherence in the coverage of the political issues and personalities. In addition, the francophone press looks at political issues with a more critical angle than does the anglophone press, as highlighted by Taras (1990) in his study of the Canadian media. For him, two very distinct worlds of journalism coexist in the country: francophone and anglophone. By focusing only on the French-language news practices variable, we could better assess the evolution of personalization. Moreover, the emphasis on French-language media sources enables an implicit comparison between the English- and French-language media in Canada, since the literature in Canada focuses primarily on the English-language press.

The year 1989 was selected as a starting point following Adatto's (2008) observation that the processes of image making and mediatization of politics were by then better controlled by both the media and political actors. This research offers a rare insight on the way personalization is treated in the Quebec francophone press.

In our quantitative content analysis and qualitative discourse analysis of all the newspaper stories, from the beginning of the race up until a week after the election, we studied framing processes based on Entman's definition (1993, 52); for him, "to frame is to select some aspects of a perceived reality and make them more salient in a communication text, in such a way as to promote a particular problem definition, causal interpretation, moral evaluation, and/or treatment recommendation for the item described."

TABLE 9.1

Political leadership races analyzed

Period of leadership race	Name of candidate	Political party	Number of stories analyzed
August 30 to December 15, 1989	Dave Barrette	NDP	6
	Phil Edmonston		1
	Audrey McLaughlin*		6
January 18 to June 23, 1990	Jean Chrétien*	LPC	105
	Sheila Copps		87
	Paul Martin		77
March 1 to June 15, 1993	Kim Campbell*	PCP	53
	Jean Charest		36
June 6, 2002, to January 25, 2003	Bill Blaikie	NDP	40
	Jack Layton*		37
	Lorne Nystrom		36
January 12 to October 14, 2003	Sheila Copps	LPC	76
	John Manley		60
	Paul Martin*		104
December 1 to March 22, 2004	Tony Clement	PCP	7
	Stephen Harper*		17
	Belinda Stronach		29
July 15 to November 17, 2005	André Boisclair*	PQ	59
	Richard Legendre		16
	Pauline Marois		42
February 1 to December 3, 2006	Stéphane Dion*	LPC	81
	Michael Ignatieff		69
	Bob Rae		43
October 20 to December 9, 2008	Michael Ignatieff*	LPC	55
	Dominic LeBlanc		28
	Bob Rae		48
Total			1,218

* Winner of the leadership race

NDP = New Democratic Party; LPC = Liberal Party of Canada; PCP = Progressive Conservative Party; PQ = Parti Québécois

Combining the two methods offered different insights into our research questions: the content analysis produced a quantitative description of the personalization processes, whereas the discourse analysis, by taking a closer look at the language used and the meaning of the descriptions, was helpful to document in the context of the gendered nature of personalization.

A total of 1,218 news stories were analyzed (see Table 9.1), focusing on one of the three frontrunners based on public opinion polls. Following Fairclough (1995) and Page (2003), we envisioned news stories both as cultural artifacts that play an active role in constructing social reality and as barometers of socio-cultural changes. News stories were coded following several criteria. First, we looked at personalization as being either based on political or on private life performances (Corner 2003; Van Zoonen 2005). Political performances are associated with political affairs practices and the institutional scene (being an MP, a business person, or an advocate for specific issues before the leadership race), whereas private life performances are related to the personal scene (appearance, family, hobbies). The visibility of either political or private life was assessed by counting the number of times it was mentioned in news stories.

A team of coders then looked more closely at the words and sentences used to portray candidates, their personality, accomplishments, experience, and private activities. Each story was read, and qualities and expressions were collected in a database. For example, expressions such as "a good family man" or "daddy's little girl" were compiled, and then the percentage of appearances was calculated to assess the weight of these descriptions in the overall portrait of the politician.

FINDINGS

As shown in Figure 9.1, and differing from theoretical expectations, personalized news did not increase during the period studied, with the exception of the Progressive Conservative Party (PCP) race, where it did go up between 1993 and 2004 (although the 1998 and 2003 races were not studied). Indeed, levels of personalization stayed mildly the same, attracting between 40 and 46 percent of the coverage from 1989 to 2003. Other stories were mostly fact-oriented, stating that the candidate had visited a specific place or given a speech, for example. Those articles were not considered personalized. In addition, media outlets treated the various political actors and races in a similar manner, in spite of diverging editorial positions.

The lower percentage of personalized stories for the 2002 NDP race could be explained by this party not traditionally attracting much of Quebec's media attention, especially when there is neither a Quebec-born nor a French-speaking contender in the race.[3] Personalization reached a peak in the Conservative Party leadership race of 2004, in which 66 percent of the coverage was personalized. It remained significant during the Parti

Figure 9.1

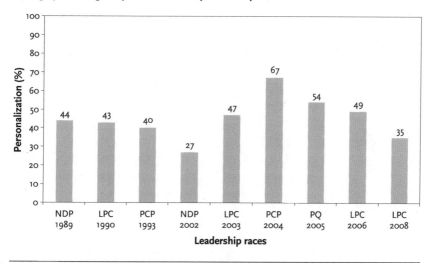

Average percentage of personalization (in articles)

Québécois (PQ) race of 2005, reaching 54 percent of the news coverage. These higher results could be explained by the amount of media attention Stronach received, as we explore in greater detail below. As for the 2005 Parti Québécois race, André Boisclair was literally chased by the media when it was rumoured that he had taken cocaine while a cabinet minister. This revelation kept journalists and editorialists speculating, as Boisclair systematically refused to give details of what he and his team dubbed "youthful indiscretions." A lot of stories emphasized that these actions were unworthy of someone aspiring to become premier of the province. Boisclair's use of cocaine was mentioned 168 times over fifty-nine articles, whereas his sexual orientation was brought up only 11 times. His apparent lack of a sense of responsibility therefore completely overshadowed the issue of his sexual orientation. Nonetheless, it can be questioned whether revelations about Boisclair's use of cocaine may have been a way for the press to tackle his homosexuality in an indirect way. This was not the case for Allison Brewer, whose sexual orientation became a key issue when she ran for the NDP's provincial leadership in New Brunswick (Everitt and Camp 2009a, 2009b). During the Liberal Party of Canada's (LPC's) races of 2006 and 2008, personalization decreased, making for only 49 and 35 percent of the coverage. These

FIGURE 9.2

Average percentage of political performances (in mentions)

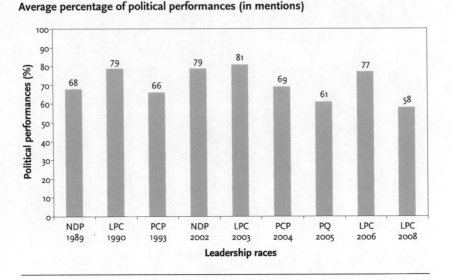

results are consistent with previous studies, which showed that both the way candidates present themselves and specific events during the campaign play a role in personalization (Kriesi 2012; Langer 2012; Lalancette and Lemarier-Saulnier 2013).

Between 1989 and 2008, what were the most important forms of personalized coverage during leadership races? As conveyed by Figure 9.2, the political experience of the candidates (past career or profession, departmental responsibilities if an MP, number of years in politics, bills passed when in power), portfolios (economic, social, or constitutional), and political endorsements are the main focus, but support for a candidate also confirms his or her skills as a leader. For example, Pauline Marois's role at the head of numerous departments and Jack Layton's years as a city councillor were highlighted as indicators of their leadership capacity.

Moreover, the political personalization was partly focused on where candidates were from and the riding they represented. For example, Audrey McLaughlin's origins were the subject of much attention (since her birthplace was Ontario yet Yukon, her electoral district), and used to assess whether she could adequately represent her constituents. In the 2003 NDP race, Nystrom was presented as the only *true* NDP candidate, since he had

FIGURE 9.3

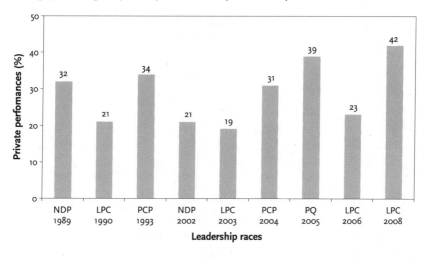

Average percentage of private performances (in mentions)

been an elected MP of the party for a long time (since 1968). Paul Martin was presented as a millionaire, but quotations from many interviews with his friends and colleagues underlined that he was "a rich man [but] not acting like one," and that he was "driving his old Jetta" and being teased about it by his friends. Our data mirror Parry-Giles's (2001) finding in the United States on this. She stresses that journalists and commentators assess the authenticity of candidates with indicators such as their relation with the state they want to represent, the absence of contradiction in their discourses, and the intentions of the candidate when running for office.

As for private life, data presented in Figure 9.3 indicate that it is not the media's main focus. The candidate's age, marital status, and family were systematically mentioned and described with many details. Their relation with their spouse, children, and pets was depicted. For example, Stéphane Dion's immense involvement when adopting his daughter was presented. Martin's wife was described as disliking politics. Kim Campbell's childlessness and divorce were mentioned many times. For almost all the candidates, and especially the women, their father is presented as having a great influence on their decision to become a politician. For example, Paul Martin Jr. was presented as the son of cabinet minister Paul Martin Sr.

For Pauline Marois, Richard Legendre, Stephen Harper, and Sheila Copps, the candidates' ability to rise above their modest origins was presented in a positive way. As for the more affluent candidates (Belinda Stronach, Paul Martin, Michael Ignatieff), the emphasis was rather put on their fortune or aristocratic background. This was particularly impressive for Stronach, as she was portrayed in press coverage as the rich, spoiled daughter of a successful businessman. Descriptions about some of the candidates' intelligence also played a role in creating a distance between the politician and citizens. It was particularly the case for Dion and Ignatieff, who were presented as being brilliant academics. Campbell's intelligence was also noticed and presented as a reason for her success. The assessment of the candidates' political and private accomplishments and setbacks was also supported with quotations from family members, friends, and colleagues, which gave credibility to these portrayals. And finally, when a woman was among the top three contenders – as was the case for the NDP in 1989, the Liberal Party in 1990 and 2003, the Progressive Conservative Party in 1993 and 2004, and the Parti Québécois in 2005 – the family issue became more significant. This leads to the question, are male and female contenders really personalized differently, as some studies seem to have demonstrated?

Echoing previous research on gender framing, our data indicate that female candidates were scrutinized more than were male candidates. As Figure 9.4 shows, 57 percent of news articles used personalization for female candidates, compared with 42 percent for male. Even when Stronach is not taken into consideration, as she could have skewed the data, 50 percent of the articles about women candidates remain personalized.

Our discourse analysis offers a different take on the quantitative data and reveals that the coverage was not gender-neutral. When we look at the various ways the political and personal life mentions are used to assess candidates, many differences are revealed. First, the viability of female candidates was often assessed negatively, as if they were not serious or real contenders. In 1990, Copps was presented as a "challenger" in the race, but not as a potential winner. Second, their political experience was often questioned, as opposed to that of their male counterparts, who were described using a more neutral tone. Marois was presented as a candidate of experience, but also as a rich snob. As noted above, this was not the case for Martin. Also, Marois's years in politics were often used to illustrate her shortcomings or lack of judgment.

Figure 9.4

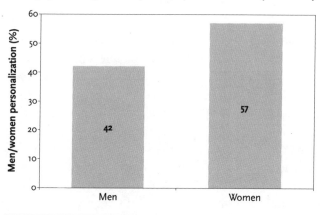

Average percentage of men/women personalization (in articles)

Third, a double standard apparently exists in the way private life was presented for female politicians. The public-private duality of the political persona seems to be more central for women, considering the frequent mentions of their personal lives. Campbell's divorce was used against her, making her unreliable, whereas Layton's and Ignatieff's divorces never came out as being problematic. Copps's divorce made headlines, and Stronach's two divorces were always included in her portrayal. The only married female candidate was Marois. However, even though she had more than twenty years of experience in politics, she was often referred to as being the wife of a former chairman of the Société générale de financement du Québec, a key state agency for economic development, instead of being "married to" or "having Mr. X for a husband." Even if women candidates mostly presented their political side to the electorate, their private lives were also regularly put up front. Maybe this was partly because they did not fit the traditional family model, whereas their opponent corresponded more closely to the quintessential family model.

This double standard applied to other criteria also. For example, Copps was presented as "just turning" fifty. For a man, this would be considered young. But for her, some would say that her "time has come" (Vastel 2003). Also, in both headlines and articles, the first name of female politicians was often used (Kim, Sheila, Belinda, Pauline), whereas male politicians were mostly called by their last name (Dion, Martin, Chrétien, Boisclair,

Layton – he only became "Jack" later on). Finally, women were often characterized as being more aggressive than their male counterparts, being presented as "fighters" and as "raising their voice," even when it was not the
case, as also noted by Gidengil and Everitt (2000).

Fourth, for women candidates, many labels reduced them to their gender and sex. Marois was presented as "the unique representative of the fem-
inine tribe to covet the leadership throne" (Boivin 2005), and McLaughlin
as "the only female candidate, but [she] doesn't put the emphasis on being a
woman to win" (Paquin 1989). It was alleged that she won because she was
a woman, an accusation also noted by Trimble and Arscott (2008). Copps
also got a gendered treatment, as the label "first female politician to ... "
stuck to her and made the headlines. "Copps became the first woman to be
candidate for the head of the Liberal Party in the history of this political
formation" (Gauthier 1990). One journalist went as far as saying that even
if she slept with all the voters in her party, it would not help change their
mind (Laporte 2003). In the 1990 race, sexist remarks about Copps (she was
called "baby" and "witch," and asked to pass the tequila; see Pelletier 1990)
were presented as the result of her pushing the male boundaries too far.
However, among all candidates, Stronach was the one with the highest rate
of privatized discourses (93 percent). She was overscrutinized, with a constant focus on her political potential or discussion about her personal life,
gender, and appearance. Descriptors such as "sexy" and "attractive" were
used forty-five times, "daddy's little girl" thirty-eight times, and "mediagenic" thirteen times in the twenty-nine articles about her candidacy. She
was also presented as having no grip on her discourse, being "a charming
woman controlled by veteran conservative organizers" (Pratte 2004). These
findings are similar to what Trimble and Everitt (2010) discovered in the
anglophone press coverage of Stronach's campaign and career. Moreover,
for the female candidates, stories constantly focused on attributes such as
their beauty, femininity, and wardrobe. In this sense, the French- and
English-language presses offered a similar coverage of the candidates with
regard to their femininity and the viability of their candidature. Both
Campbell's and Stronach's blonde hair was also put forward. These are seen
as "discursive techniques" aimed at domesticating and feminizing women
leaders (Trimble and Arscott 2008). For men, their look was seldom mentioned, and stories generally presented their past experience, age, and family
status more neutrally.

Political Communication in Canada

Following Vliegenthart, Boomgaarden, and Boumans (2011, 108), we think that personalization is a "multi-faceted phenomenon," and that we have only captured a small part of it. We must therefore ask whether personalization is a phenomenon more closely associated with the personality of the leaders, and to their relationship with the media, or whether it's a general tendency.

Some leaders are more media-friendly than others, and play with celebrity culture codes effortlessly (the late Jack Layton was one of them, as opposed to Stephen Harper; but see Chapter 4). We also suspect that the phenomenon could be more acute during electoral campaigns than during leadership races, as stakes are higher and issues more diverse. Our focus on French-language newspapers may also be questioned by asking whether personalization is more common in this type of media. A systematic comparison could be made with English Canadian media to document the nature of the phenomenon on a larger scale. However, it would be interesting to see whether television, popular press (entertainment and lifestyle magazines, for example), and new media (social networks) offer a more personalized coverage of politics. As Grabe and Bucy (2009) observed, the image of the politician has considerable impact on voters, as they evaluate candidates based on their perceived personality traits and communication skills. Moreover, considering the growing influence of new forms of media, the visual aspects of politics may become even more central. In that sense, the role of communication advisers should also be taken into consideration. In addition, the role of parties in the assessment of candidates might be further investigated.

What are the implications of the personalization phenomenon for the future of politics in Canada? We suspect that there will be a greater focus on the political persona, as qualitative data has shown, and that more questions will be raised concerning the authenticity, honesty, and character of candidates, as seen in our study and also discussed by academics such as Ankersmit (2002), Bashevkin (2009a), Van Zoonen and Holtz-Bacha (2000), and Frau-Meigs (2001). Furthermore, our case study highlights the way female candidates are framed by gender (Van Acker 2003; Trimble, Treiberg, and Girard 2010), and that the glass ceiling could perhaps also be one of media representations. In a similar way, the attention on Boisclair's sexual orientation was put forward through his cocaine use. His apparent lack of a sense of responsibility therefore completely overshadowed the issue of his sexual orientation. Nonetheless, it can be questioned whether revelations about Boisclair's use of cocaine may have been a way for the press to tackle his homosexuality in an indirect way. This raises questions about the coverage of "unconventional" persona and the implicit heterosexual norm of politics. Coverage of women or gay politicians stressing their uncommon path could lead to questioning the legitimacy of their candidatures. At the same time, it could also lead people to choose a different career in order to avoid being stigmatized in the media.

The role private life plays in news stories points to the impacts of what researchers call the "privatization" of politics, characterized by the presentation of candidates in their private roles and environments (Langer 2012). This privatization could be considered as offering greater access and thus proximity to political leaders, to the human being behind the politician. For example, that Dion has an adopted daughter was presented positively, and could influence the way voters see him and assess his potential as a leader. Similarly, that some women candidates were divorced was presented negatively, as an indicator of unreliability.

Likely related with celebrity politics (Pease and Brewer 2008; Thrall et al. 2008; Street 2011) and a new aesthetic of politics (Corner and Pels 2003), this increased interest in the private life of politicians could modify the way we see politics in the near future. Should this be alarming? Maybe not. Following Dakhlia (2009, 2010), we think that the contemporary assimilation of politics into more spectacular practices invites us to rethink democratic representation, and its highly aesthetical dimension.

Finally, how does personalization influence citizens? Could it explain the current decline in political participation, discussed among others in the media malaise literature? Many hypotheses may be issued about the relationship between the two phenomena. As Bittner (2011) noted in her research, leaders do matter, and voters evaluate leaders' traits based on character and competence. Also, as Karvonen (2010) observed, a person and his or her political party are not opposites. People already interested in politics will care about both, and indifferent citizens will care about neither parties nor candidates. To further assess these potential changes and better understand the changing nature of political communication, new analytical tools need to be developed.

NOTES

1 The research project from which this chapter is drawn was supported by a grant from Fonds institutionnel de recherche (FIR) in 2010-12, given by the Université du Québec à Trois-Rivières. The project examines the newspaper coverage of male and female leadership races in the context of spectacularized and personalized politics. This chapter is a revised version of a paper presented at the Canadian Political Science Association conference held in Edmonton in June 2012. The authors thank the anonymous peer reviewers for their thoughtful comments, as well as the editors for their great work.

2 Some research has been done about personalization in talk shows (Van Zoonen and Holtz-Bacha 2000) and current affairs programs (MacDonald 1998). Other research documents the effects of the personalization on voters' perceptions of politicians and parties (Kaase 1994), as well as personalization as a process to get more voters (Nielson 2012).

3 This is why the 1995 federal NDP leadership race was excluded from our study.

10

The Mass Media and Welfare Policy Framing: A Study in Policy Definition

Adam Mahon, Andrea Lawlor, and Stuart Soroka

Political communication – particularly the study of media content – has much to tell scholars about public policy and public attitudes toward policy. The existing policy literature reflects two interpretations of the mass media in the policy process. Some work focuses on the media as simply a conduit for information about the policy process; other work focuses on the media as a cause of policy change (see Kingdon 1984; Baumgartner and Jones 1993, 1994; Sabatier and Jenkins-Smith 1993; Shanahan et al. 2008). In short, the media can be viewed as driving public opinion about policy, or simply reflecting it. Either way, analyzing the media can help scholars better understand the state of public opinion on policy and enhance our understanding of the policy process.

Consider first the growing literature suggesting that the media are a cause of policy change, either directly or indirectly (through changing public opinion, which then affects policy). Work on *agenda setting* points to the media as a key source in ordering up policy issues for debate by both legislators and public opinion (Cohen 1963; Lang and Lang 1966; McCombs and Shaw 1972; Cobb and Elder 1983; Nelson 1984; see also Elisabeth Gidengil's Chapter 8 in this book). The media help determine which issues are addressed by the public and by policy makers. The literature on *framing* focuses on how variations in the presentation of information can help cultivate particular understandings of issues, largely by selecting (and deselecting) certain aspects of stories (Edelman 1985; Iyengar and Kinder 1987; Iyengar 1990, 1991; Entman 1993; Scheufele and Tewksbury 2007). This work suggests that the way in which a policy is framed in the media can have a sizable impact on public outlook; in fact, the act of making even small changes to the presentation of an issue or event can produce substantial changes of opinion on that issue (see especially Chong and Druckman 2007, 104). The media necessarily play a large role in creating and distributing these shared frames – although they are not the only source of political information, nor the only source of shared frames around political issues, the media do create the "universe of discourse" that citizens tend to consume (Blumer 1946).

As Daniel J. Paré and Susan Delacourt discuss in Chapter 7, work on the study of media gatekeeping explores, in part, the relationship between the media and policy, with a focus on the way in which new production and norms condition the issues that the media tend to select for discussion. News editors have been shown to be highly subjective in the stories they let through to publication (White 1950, 386; see also Lewin 1947; Snider 1967; Berkowitz and Adams 1990). McNelly (1959) suggests that the same is true of reporters and the stories they decide to pursue. Pamela Shoemaker's (1991) work has been perhaps the most influential in advancing theories of gatekeeping. Shoemaker outlines the way in which the attitudes and preferences of journalists and editors systematically influence news output; she also reviews the impacts of the routines and practices of news staff and news media ownership on content (also see studies by Gieber 1964; Epstein 1973; Dimmick 1974). What arises, she and others (e.g., Gans 1979) argue, is a hierarchy among news stories presented to the public and a powerful role for the media to influence the content and strength of public opinion on a policy issue.

There is, in sum, a considerable body of work suggesting that the mass media can affect public policy either directly or indirectly. That said, the media can also simply reflect what is going on. Indeed, media content may capture the stasis or change in the policy process better than any other actor in the policy process. This is largely because the media are the most likely, and sometimes the only, source of national political information. As such, they "provide the best – and only – easily available approximation of ever-changing political realities" (McCombs and Shaw 1972, 185). A principle function of the mass media, after all, is to act as a conduit between publics and policy makers.

Of course, in reflecting policy, the media can influence the way people perceive issues. As noted, the media can shape public understanding of policy simply by defining an issue in particular terms or by choosing specific language to illustrate public policy. This influence can be intentional or not; it can be effective or not. The point here is that media *effects* need not be the source of value for media content in studies of public policy. Regardless, the mass media's reflection of the policy process can be relevant to the way we understand policy.

Put more succinctly: as a contributor, the media can advance a particular policy agenda and affect policy directly; but as a conduit in the policy process, the media transmit information back and forth across actors, and in so doing can also capture the shifting nature of the debate (for more on this distinction, see especially Shanahan et al. 2008, 115). Using the media in analyses of public policy thus need not be focused on situating the mass media in the policy process itself. Authors such as Baumgartner and Jones (1994) and Cobb and Elder (1983) note, for instance, that the media's reflection of public policy tells us much about the broader policy process that we may not otherwise glean from studies that focus exclusively on individual policies. A more typical in-depth look at specific policies or networks of actors gives audiences information about particular policies and actors, but may fail to give insight into the larger context in which those policies and actors are situated. As such, more typical approaches to policy studies may miss out on the high-level view of the policy process, leaving a gap to fill in policy research, one that can be filled (at least partially) by media analysis. The point is that the role the media play in the policy process may not be critical to its value for scholars of policy – what may be central is the frequent capturing in media coverage of the content, the nature, the tone, or the participants in policy making. As Murray (2007, 526) notes, "No other independent institution ... reaches as many citizens daily"; therefore, the way that the media characterize a policy debate is relevant to a broader understanding of how the public perceives that domain. Media data thus become a valuable resource for those interested in policy making, which is to say, political communication can make a valuable contribution to the study of public policy.

Case Study

If public opinion surveys are any indication, Canadians have a rather strong distaste for welfare (see Harell, Soroka, and Mahon 2008; Soroka and Robertson 2010). However, relatively little work explores the root causes of the resentment toward welfare (or the seemingly negative opinions of welfare recipients) of many Canadians. The same is not true in the US context, where much emphasis has been placed on the role the mainstream media play in shaping opinions toward welfare, and it is widely held that the media's traditional approach to welfare has been decidedly negative. That is, conventional wisdom holds that Americans hate welfare in part because of how the media frame issues related to social assistance. The Canadian literature has not yet tested this possibility.

A first step, taken here, is to examine the nature of Canadian media coverage of welfare. We select welfare as the topic for our case study because of its continuous relevance to Canada's broader social-policy landscape. Welfare policy is clearly a positional issue that tends to elicit strong opinions from the public; furthermore, we hypothesize that its prominence in the public debate lends itself to various frames for discussion.

First, what is framing? Iyengar (1997, 214) identifies framing as a form of subtle influence, which derives not from the amount of news presented on a particular issue but from "the manner of presentation of news on that issue." In short, for the majority of the public, political issues are defined by media coverage of those issues; the manner in which such issues are presented thus has a considerable impact on public opinion. Iyengar's (1990) experimental study on social assistance policy is particularly relevant here. Iyengar finds that framing issues related to poverty or social assistance policies in either societal or personal terms (e.g., widespread economic difficulties versus particular instances of individuals living in poverty) influence participants' views of whether poverty is a consequence of societal or of personal factors. Those causal attributions then affect participants' opinions on social assistance policies, either weakening or strengthening support for welfare. In sum, Iyengar finds that "the framing of political issues is a powerful form of social control that circumscribes ... debate over public policy" (38).

Frames embody a narrative that weaves together a series of concepts to convey a particular meaning or outcome (Hertog and McLeod 2001). Central to our analysis here is the idea that frames are systematically linked to a particular language, which is to say, to words. Apart from the occasional photo, the manner in which newspapers present the news, for instance, is quite obviously contingent entirely on the use of words. (The same is true for television news, of course, though in this case pictures can matter to framing as well). For example, a newspaper article describing a social assistance program using the term "welfare" may include other terms such as "need," "collective responsibility," and "societal values." Conversely, welfare may be associated with terms such as "fraud," "abuse," and "dependency." The framing of a policy item in a negative manner may create or reinforce negative attitudes toward the policy.

In the case of social assistance programs, even the name of the policy may matter. "Welfare" is one name ascribed to the policy domain, and the selective use of this term in media coverage of social assistance policies may serve to "reinforce a particular moral universe" (Mirchandani and Chan 2007, 25), the most clear expression of which may be public opposition to social assistance programs. For instance, the effects of media framing of welfare policies are evident in work focused on responses to survey questions in this domain. In both the United States and Canada, survey respondents are generally much more likely to support cash assistance to the poor if the question avoids the term "welfare" (Iyengar 1991; Harell, Soroka, and Mahon 2008). To many people, "welfare" elicits reminders of "deadbeats, fraud, and entangled bureaucracies; and ... serious disincentives for employment" (Shapiro and Young 1989, 73). Earlier work undertaken in the American context has suggested that an additional explanation for public distaste for "welfare" may be partly explained by racial bias, as "attitudes of the mass public toward welfare spending are closely tied to racial attitudes" (G.C. Wright 1977, 728). Race is likely part of the story in Canada – recent work suggests that perceptions of Aboriginal communities are a factor in Canadians' support for welfare policy (for instance, Harell, Soroka, and Ladner 2012; Banting, Soroka, and Koning 2013). But other factors may be equally impactful, such as views related to the deservingness of candidates. We thus far have only a vague sense of why the word "welfare" elicits relatively low levels of support in the Canadian context. But as survey results make clear, the words used to describe policies can make a big difference to media frames, and to policy support.

METHOD

We explore media framing of welfare using a dataset consisting of 9,977 articles published by nine newspapers and the Canadian Press newswire service between January 1, 2000, and December 31, 2006. These ten media outlets belong to three categories: the Canadian Press, Canada's most substantial newswire service; the three most widely circulated English-language newspapers in the country – the *National Post, Globe and Mail,* and *Toronto Star* – and six of the country's most heavily circulated local newspapers – the *Montreal Gazette, Ottawa Citizen, Hamilton Spectator, Winnipeg Free Press, Edmonton Journal,* and *Vancouver Sun.*[1] All articles are drawn from the Factiva full-text database. Articles from the six local newspapers are included only in the final year of the sample (as those newspapers aren't reliably indexed before that time), whereas the dataset includes articles from the three national newspapers and the Canadian Press for the entire seven-year period under examination.

Among the nine newspapers in the sample, five were owned by CanWest Global throughout most of the period (*National Post, Gazette, Edmonton Journal, Ottawa Citizen,* and *Vancouver Sun*), two were part of Torstar (*Toronto Star* and *Hamilton Spectator*), one was independently owned (*Winnipeg Free Press*), and one was the property of the former Bell Globemedia (*Globe and Mail*). Although the nine newspapers in the sample account for 38.7 percent of the average weekly circulation among Canadian newspapers (based on 2007 data), certain owners are either over- or under-represented. For example, although newspapers owned by CanWest (now Postmedia Network) – the chain whose newspapers enjoy the highest combined levels of circulation – represent 50 percent of the newspapers in the sample, these papers account for only 27.4 percent of average weekly newspaper circulation in Canada. Notably absent from the papers included in the sample are any of those belonging to the Quebecor/Sun Media chain: news published by Quebecor account for 21 percent of average weekly newspaper circulation, which ranks this company second among Canadian newspaper owners.[2]

Having determined the period to be examined, and the newspapers to be included in the sample, a detailed search query was designed to capture all articles containing the word "welfare," while excluding those articles in which it was used in the contexts of the phrases "animal welfare," "child welfare," "corporate welfare," "health and welfare," and "welfare of..." In addition to articles containing the word, the search query included several terms that together capture the official names of all major provincial and federal social assistance programs during the period under analysis. These terms include "Canada Assistance Plan," "public assistance," "social assistance," "income support," "income assistance," and "welfare assistance," as well as the terms "workfare," "Ontario works," and "Alberta works" – used in mentions of provincial workfare programs. Finally, terms related to public housing, a prominent aspect of social assistance policies in Canada, were included (e.g., "public housing," "social housing," and "subsidized housing").

This search query yielded initial results of 13,525 articles, though we reduced the overall number by dropping non-pertinent and duplicate articles, as well as those appearing in non-news sections of the newspaper, such as Entertainment. In the end, the working database included 9,976 articles. (See Table 10.1 for the distribution of articles across sources.)

TABLE 10.1

Articles in dataset, by newspaper, by year

	2000	2001	2002	2003	2004	2005	2006	Total	% total
Canadian Press	231	400	444	266	167	138	299	1,945	19.5
Edmonton Journal	—	—	—	—	—	—	193	193	1.9
Globe and Mail	353	286	308	260	263	365	382	2,217	22.2
Hamilton Spectator	—	—	—	—	—	—	172	172	1.7
Montreal Gazette	—	—	—	—	—	—	202	202	2
National Post	165	334	334	290	291	219	205	1,838	18.4
Ottawa Citizen	—	—	—	—	—	—	290	290	2.9
Toronto Star	470	158	550	0	431	472	549	2,630	26.4
Vancouver Sun	—	—	—	—	—	—	319	319	3.2
Winnipeg Free Press	—	—	—	—	—	—	170	170	1.7
Total	1,219	1,178	1,636	816	1,152	1,194	2,781	9,976	—
% total	12.2	11.8	16.4	8.2	11.5	12	27.9	100	100

These data are analyzed here using Provalis Research's QDA Miner qualitative analysis software. Our interest is in the relative frequency of frames in welfare coverage; we extracted those frames using a categorical dictionary that was inductively extracted from the dataset itself.

Influenced by automated frame analysis techniques pioneered by Mark Miller (Miller and Riechert 1994; Miller 1997), defining welfare frames in the media began with generating a list of frequencies for all words found in the headline and lead paragraph of every article in the dataset. Content analysis software typically features an exclusion process or a user-defined "exclusion list" that ignores commonly used words when word count frequencies are generated. Using a slightly modified version of WordStat's default exclusion list, the exclusion process was used to remove words with little interpretive or descriptive value, such as pronouns and conjunctions. WordStat also performs stemming and lemmatization processes, which reduce various word forms and tenses to their common root, transforming pluralized occurrences of a particular word to that word's singular root, and replacing past-tense verbs with their present tense, and so on. That is, with the stemming and lemmatization feature enabled, WordStat counts each occurrence of the words "stemmed," "stemming," and "stems" as occurrences of the word "stem."

A two-pronged approach was then used to determine the terms that would appear in the categorization dictionary. First, we focused on just the first 600 most frequently occurring words (of the 85,000 unique words counted by WordStat); we then excluded words that did not appear to have any particular meaning where welfare framing was concerned (e.g., "people" or "live"), along with proper nouns. The remaining 108 terms were used in our framing dictionary. We then repeated the process for phrases of two to five words. The initial count yielded a total of 16,155 phrases; our filter reduced that number to 167 relevant phrases, added to the framing dictionary.

The final state was to conduct a search for the 275 words and phrases in the framing dictionary, employing a hierarchical clustering analysis. In general terms, clustering is a procedure by which observations within a corpus are grouped together, using a similarity coefficient, and according to a set of criteria established by the researcher – in this case, the co-occurrence of the terms in the framing dictionary. Our analysis looked at the co-occurrence of terms within the same article, using the Jaccard coefficient as a similarity measure. The Jaccard coefficient for two words is calculated using a rather simple formula, expressed as $J = a / (a + b + c)$, wherein a represents cases in which both words and phrases occur, whereas b and c both represent cases in which only one of the two items appears. WordStat is able to present the resulting clusters in the form of a dendrogram, or tree graph, wherein words form the leaves, the linkages between those words form the branches, and the trunk of the tree is the point at which all branches meet. This form of clustering allows the user to indicate the specific number of clusters to be included. To allow for diversity in the frames that emerged (though not too much diversity), the number of clusters permitted was set as being no less than six and no more than twelve, whereas the exact number of frames to be used in the final analysis were allowed to emerge naturally from the text.[3]

FINDINGS

Results of the hierarchical clustering analysis are illustrated in Figure 10.1. There are eleven clusters in total, suggesting eleven relatively clear frames in our sample of welfare articles:

- *Aboriginal*: Issues relating to Aboriginal peoples and communities.
- *Crime/Victim*: Issues relating to crime and the victims of crime.
- *Economic*: Issues related to economics, finance, and taxes.
- *Health*: Issues related to health and health care.

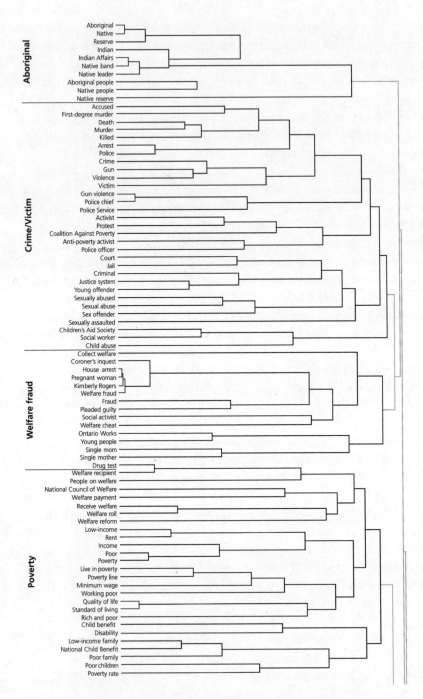

FIGURE 10.1   **Clustered terms within sample of welfare articles**

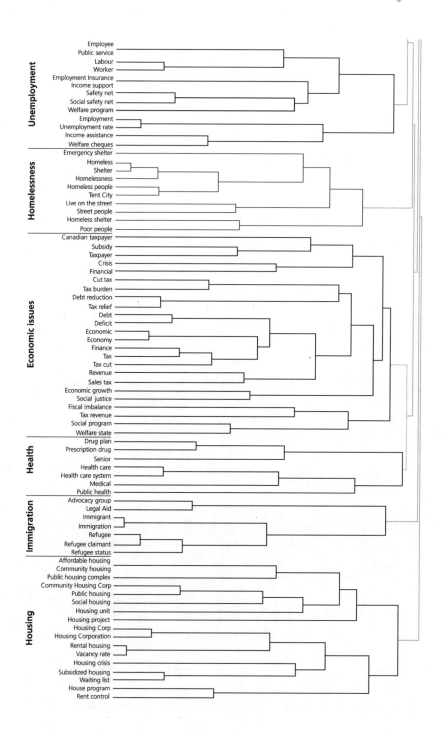

- *Homelessness*: Issues relating to homelessness.
- *Housing*: Issues relating to subsidized and social housing.
- *Immigration*: Issues relating to immigration, immigrants, and refugees.
- *Poverty*: Issues related to poverty generally, and child poverty specifically.
- *Protest*: Issues relating to protest; more specifically, anti-poverty activism.[4]
- *Unemployment*: Issues relating to unemployment and Employment Insurance.
- *Welfare fraud*: Issues relating to welfare fraud and policy approaches to fraud.

Having identified these frames, we then recounted the occurrence of all the terms, grouped by frame. Articles were identified as including a frame if they included at least two terms within a framing category (the eleven categories, and words). The resulting frequency of frames in welfare stories are shown in Table 10.2.

Table 10.2 shows the percentage of stories using each of the eleven frames. Note first that 90.3 percent of the articles in the dataset were found to represent at least one of the eleven frames selected. That is, among all articles relating to welfare printed within all newspapers in this sample, only 9.7 percent of articles did not address the issues represented by at least one of the eleven frames. We see this as evidence that the frames identified above capture a good deal of what is going on in media coverage of welfare.

We placed no restrictions on the number of frames that can be evident in a single article (meaning that, looking down columns of Table 10.2, the percentages need not sum to 100). Indeed, 58.3 percent of the articles contain two or more of the eleven frames, with the mean number of frames represented in all articles at 1.92. Only 9 percent of the articles in the dataset included two or fewer dictionary terms; the mean number of dictionary terms represented in all articles was much higher at 13.62. Again, we see this as evidence of the comprehensiveness of our frame categories. And note that to the extent that each frame word serves to invoke specific associations (for the public or for policy makers) in a context linking them to discourse on welfare, a narrative that contains on balance over 13.0 of such cuing words seems likely to have quite a strong effect on perceptions.

How do frames fare vis-à-vis one another? The most frequently represented frame across time (and across all newspapers also) is the economic issues frame, evident in 46.3 percent of articles. The words associated with

TABLE 10.2

Frames (frequency and percentage), all newspapers and Canadian Press, 2000-6

	2000		2001		2002		2003		2004		2005		2006		Total	
	n	%	n	%	n	%	n	%	n	%	n	%	n	%	n	%
Aboriginal	81	6.6	93	7.9	96	5.9	55	6.7	64	5.6	108	9.0	194	7.0	691	6.9
Crime/Victim	398	32.6	431	36.6	641	39.2	259	31.7	388	33.7	422	35.3	954	34.3	3,493	35.0
Economic	581	47.7	579	49.2	781	47.7	424	52.0	536	46.5	521	43.6	1199	43.1	4,621	46.3
Poverty	387	31.7	368	31.2	502	30.7	222	27.2	366	31.8	348	29.1	927	33.3	3,120	31.3
Health	127	10.4	132	11.2	251	15.3	131	16.1	198	17.2	109	9.1	302	10.9	1,250	12.5
Homelessness	128	10.5	83	7.0	223	13.6	69	8.5	137	11.9	104	8.7	299	10.8	1,043	10.5
Housing	132	10.8	103	8.7	255	15.6	69	8.5	156	13.5	200	16.8	408	14.7	1,323	13.3
Immigration	58	4.8	73	6.2	94	5.7	59	7.2	106	9.2	91	7.6	178	6.4	659	6.6
Protest	75	6.2	78	6.6	88	5.4	43	5.3	52	4.5	44	3.7	98	3.5	478	4.8
Unemployment	259	21.2	226	19.2	330	20.2	173	21.2	210	18.2	212	17.8	554	19.9	1,964	19.7
Welfare fraud	63	5.2	84	7.1	204	12.5	45	5.5	80	6.9	79	6.6	159	5.7	714	7.2
Total	1,219	12.2	1,178	11.8	1,636	16.4	816	8.2	1,152	11.5	1,194	12.0	2,781	27.9	9,976	100.0

this frame include tax policy, debt and deficit, the fiscal imbalance, and economic growth. (See Figure 10.1 for a full list of terms.) These are relatively common terms in news reports, perhaps, about welfare or not. But it is important that welfare policy need not necessarily be connected to a discussion of macroeconomics and tax policy. After all, more than half the articles in our sample do not mention economic issues keywords. We thus see the prevalence of the economic frame as a potentially meaningful indication of the way in which welfare policy is depicted in public debate in Canada.

It is perhaps more striking, and likely more meaningful as well, that the second most-represented frame in welfare stories is the crime frame – evident in 35 percent of all stories. The prominence of crime fits with the existing literature cited above on the mainstream media's approach to reporting on issues related to welfare in both Canada and the United States. To date, however, the Canadian literature has provided only a very partial picture of the prominence of crime in public discussions of welfare. Data in Table 10.2 thus provide new and much more comprehensive evidence of the phenomenon.

The third most frequent frame among the cases in the dataset was the poverty frame (31.3 percent of cases). The frequency of this frame is unsurprising given that this frame is composed of terms associated generally with poverty but also specifically with "child poverty," an issue that has been quite salient on both the media and political agendas in North America since the 1970s and 1980s (see Nelson 1984). That the poverty frame is less apparent than either economic issues, or crime, however, is striking. Poverty, the central issue in welfare policy, is less apparent in articles dealing with welfare than is crime.

When all newspapers are considered together, there is a rather sizable gap between the three most frequently represented frames and the remaining eight, as the fourth most frequently occurring frame is the unemployment frame, which appears in only 19.7 percent of cases. Following the unemployment frame in frequency are the frames of housing (13.3 percent), health (12.5 percent), homelessness (10.5 percent), welfare fraud (7.2 percent), Aboriginal issues (6.9 percent), immigration (6.6 percent), and protest (4.8 percent). These issues are clearly not as prominent in media coverage of welfare. That said, we do not want to minimize their potential significance. Recent Canadian contributions on the relationship between immigration and support for welfare (Harell et al. 2012) and on the (negative) impact that attitudes about Aboriginals can have on support for welfare (e.g., Harell, Soroka, and Ladner 2012; Banting, Soroka, and Koning 2013) have highlighted these aspects of the debate. Taken together, these frames occur in more than 10 percent of welfare articles in Canadian newspapers. And nearly one in ten articles on welfare includes more than two words in our welfare fraud frame. Clearly, even at relatively low levels overall, these frames can be of real importance in the way Canadians apprehend welfare issues, and to the nature and tone of public debate on welfare policy.

Although the data in Table 10.2 show fluctuations in the proportion of coverage individual frames receive over time, none of these fluctuations is of a magnitude to signal a major shift in the media agenda on welfare. The period under examination is relatively short, however, and there is good reason to believe that changing frames – particularly for an issue of only moderate salience – occurs rather slowly over time. US studies suggest that shifts within the media's welfare agenda tend to coincide with movements toward welfare reform, which in turn tend to coincide with shifts in political leadership (Gilens 1999). This being the case, a major shift in the media's welfare agenda during this period is perhaps not to be expected: over most of the period under examination, the same party held power at the federal level of government.

We do not show differences in coverage across newspapers here, but it is worth noting that some exist (though see Mahon 2009 for full results). For instance, although in each of the major national newspapers the proportion of coverage given to economic issues exceeded that of other issues, the same did not hold for the smaller newspapers in the sample. In the *Edmonton Journal* and *Ottawa Citizen,* crime was the most frequently occurring frame, whereas in the *Hamilton Spectator,* stories representing the poverty frame exceeded those representing economic issues by a substantial margin. And both the crime and economic issues frames are represented in exactly the same proportion of stories printed in the *Winnipeg Free Press.* The somewhat higher emphasis on local crime stories in regional newspapers has been well documented elsewhere (see Reber and Chang 2000); these results suggest that welfare coverage may in some newspapers be coloured by this focus on crime – perhaps with important consequences where public attitudes toward welfare are concerned. This is only conjecture at this stage, admittedly. We cannot yet easily connect our findings on media coverage to variations in public support for welfare. However, we do now have a general sense for the frames more prevalent in Canadian media coverage of welfare over a six-year period.

Political Communication in Canada

Much can be learned from media content about both public opinion and policy development. This chapter has illustrated this by focusing on media coverage of the welfare policy domain. There is good reason to believe that the Canadian print media's coverage of issues related to welfare in Canada is shaped by the political agenda. That is, those welfare issues that factor largely in the political discourse – however relevant or not in reality – likely affect the media's portrayal of welfare. There is no better example of this phenomenon than the manufacture, by the Ontario Progressive Conservative government in the late 1990s and early 2000s, of a crisis of welfare fraud. Empirically, no such crisis could be said to exist, but the media effectively extended the provincial government's crusade against welfare fraud to the pages of Canada's most widely circulated daily newspapers.

But the media do not just reflect the political agenda. Public concerns, and journalists' concerns, matter to the nature of media content as well. We should thus view media content as some combination of these factors. The media reflect, and in some cases affect, the public and policy makers. For a wide range of issues, then, media content provides a lens through which to better understand the structure of current debates on public policy.

We have, of course, only started to understand the welfare policy debate here. Yet the preceding expository analysis is telling. There is a proliferation of coverage of crime and macroeconomic issues in media content on welfare – in both cases, more than there is discussion of poverty or homelessness. There are also steady levels of coverage using the fraud, immigration, and Aboriginal frames. This distribution of frames point toward a decidedly negative bent to media coverage of welfare during the period examined. Coverage of such frames likely serves to problematize social assistance, casting it as a policy domain defined principally by dependency and abuse, while drawing focus from discussions of those issues to which these policies have emerged as a necessary response – namely poverty and homelessness.

Media decisions to emphasize some aspects of particular public policies over others, whether deliberate or not, shape media consumers' perceived social realities. And if media decisions are such that they systemically marginalize particular issues, then those issues are unlikely to appear in public debate. The capacity of citizens to develop informed opinions on the successes and failures of a wide range of policies is contingent in large part on media content. It follows that the nature of media coverage can be of fundamental importance to public attitudes about policy, and to policy itself.

What do the results here have to say about the themes addressed throughout this volume? Broadly, the chapters speak to issues involving the tools used by political actors and citizens to communicate, the extent to which the current political communication environment produces informed and engaged citizens, and the resulting implications for the nature and quality of representative democracy in Canada. The case examined above – welfare coverage from 2000 to 2006 – is relatively narrow, to be sure. But our work clearly has broad implications. The nature of political communication on policy issues should be viewed not just in terms of which issues receive attention but also in terms of the ways in which those issues are framed – indeed, how the language is used when those issues are discussed in the public sphere. Words (and pictures) do matter to the way in which publics and policy makers understand policy issues; they likely matter to the media's role in producing well-informed, engaged citizens as well. There is ample evidence in this book that the mass media play a critical role in representative democracy, and more specifically, that the quality of democracy is partly contingent on the quality of information in media content. It follows that the framing of policy issues in the media may be of central importance not just to how we understand policy but also to how we understand the representative democratic process.

Where findings on welfare coverage are concerned, our findings point to biases in media coverage – biases that may help explain why public support for "welfare" tends to be relatively low. The next step is to more directly connect trends in media content with trends in public opinion and policy. A small but growing body of work uses media content to better understand the policy process. We have shown here the potential for inductive content-analytic techniques in this regard. Future work will, we hope, take advantage of the methods described here, which can be readily applied to other policy domains. Doing so has the potential to add much to what we understand not just about the mass media but about public policy as well.

NOTES

1 Only English-language newspapers were selected for analysis, as the method by which frames of welfare coverage were inductively extracted from the dataset itself required that analyzed texts share a common language. Although roughly similar frames may emerge within news coverage published in both French and English, the frames extracted would not be entirely populated by terms sharing identical literal meanings. In short, separate sets of English- and French-language frames

would result, wherein across newspapers published in the same language, the basis for comparison is much stronger.

2   Quebecor newspapers were excluded because Factiva does not contain complete archives of the content of Quebecor newspapers.

3   At this stage, further amendment of the dictionary was necessary, as the presence of several exceedingly broad, and less frequently occurring, terms disrupted or altered what otherwise seemed to be natural clusters. If the meaning of a word is too broad, if it possesses multiple meanings, or if its potential uses are exceptionally versatile, it is not a suitable candidate for clustering. For example, the word "fight" occurs with great frequency among the articles in the dataset. However, in one context, the word "fight" can describe a physical altercation; in another, it may be used to describe a legal battle. In yet another context, a person may be "fighting an addiction," or "fighting for their life."

4   The protest cluster was manually extracted from the crime/victim frame. Although the terms in both frames clustered together naturally – and for obvious reasons – they were separated because the terms in the protest frame were thought to represent a unique phenomenon worthy of measuring independently.

PART 3

Political Communication and Canadian Citizens

11

Opportunities Missed: Non-Profit Public Communication and Advocacy in Canada

Georgina C. Grosenick

Non-profit organizations – those organizations separate from the corporate and government realms that serve a social or mutual benefit – are important political communicators. Although we often associate non-profit political communication activities with special interest and social movement organizations (e.g., Greenpeace), it is imperative to recognize that service and charity organizations (e.g., a homeless shelter or the Canadian Cancer Society), that make up the larger percentage of the non-profit sector are also important advocates in debates on social and political issues. Effective advocacy, for any non-profit organization, is contingent on it communicating with citizens and decision makers external to the organization and presenting its message in a way that encourages them to think about issues differently and to take action to resolve those issues. Many non-profit organizations are ineffective or inactive advocates. Structural and capacity issues impact their efforts; however, they have also not yet strategically maximized the communication opportunities available to them to reach these needed audiences and to affect social and political change.

There are approximately 161,000 non-profit organizations in Canada (Imagine Canada 2006). The non-profit sector (also commonly referred to as the third sector, the voluntary sector, or the non-governmental organization sector) is extremely diverse, but organizations within it share many commonalities, such as a formal organizational structure; operations separate from the government and business sectors; the reinvestment of earnings back into the organization; self-governing and voluntary practices; and the desire to achieve a collectively determined normative goal or mission (Febbraro, Hall, and Parmegiani 1999; Salamon et al. 1999). The perception of the collective nature and independence of non-profit organizations from government and commercial influence positions them as essential actors in the political and legislative process, allowing for "wider notions of public good or public interest" (E.J. Reid 1999, 291) to be incorporated into the public debate (in reality, many non-profits are heavily reliant on the government and corporate sectors for their sustainability). Further, including non-profit participation in the debate allows for marginalized or otherwise minimized voices in society to be included (Kramer 1981; Salamon 1987; Eikenberry and Kluver 2004). Yet non-profits are largely absent from these policy debates.

Funding pressures, structural inhibitors, and regulations influence the willingness and ability of many non-profits to communicate their views about legislative and policy issues. First, a heavy reliance on government funding, which accounts for almost 50 percent of non-profit revenues (Imagine Canada 2006), self-censors many organizations from speaking out against government policy lest they be perceived to be biting the hand that feed them (Pross and Webb 2002; Schmid, Bar, and Nirel 2008). Second, policy development structures that value individual citizen voices over the non-profit organizational voices (Phillips 2001, 2006, 2007) and the increasing concentration of power at the executive, elected level of Parliament (Aucoin 2008; Savoie 1999b) further exclude the sector from the policy process. Finally, legislatively, the Income Tax Act sets limits on the "political activity" efforts of charity organizations and threatens the revocation of the organization's charity status for non-compliance. "Political activity" is defined as any activity that "furthers the interests of a particular political party; or supports a political party or candidate for public office; or retains, opposes, or changes the law, policy or decision of any level of government in Canada or a foreign country" (Canada Revenue Agency 2003).

Relationships with governments also impact non-profit political communication. Susan Phillips (2007) observed that an "advocacy chill" pervades government-sector relations, discouraging non-profits from commenting on or critiquing legislation and programs. Today, the situation is more pronounced. In 2012, the Conservative government cancelled funding to a number of non-profit environmental advocacy groups and publicly questioned their involvement in policy development. Then minister of natural resources Joe Oliver suggested that these groups speaking out were "seeking to hijack the regulatory system" (Imagine Canada 2012). These actions and comments reinforce perceptions within the non-profit sector that they are not welcome partners in the policy development process (ibid.). What results are "practices of restraint" for the majority of the non-profit sector

(Webb 2000, 251). Few charity and service non-profit organizations dedicate resources to political communication activities (Berry 2003; Phillips 2007), and direct lobbying remains a low to moderate level of activity for most organizations (Schmid, Bar, and Nirel 2008).

To accomplish their political goals, non-profit organizations must find other ways to participate in the policy process. Agenda-setting theory offers insight into an alternative means of participation. First introduced by McCombs and Shaw (1972) as an extension of the media effects model, agenda-setting theory illustrates the importance of the mainstream media for prioritizing, and hence directing attention to, issues among citizens and decision makers. Early adaptations present a media-centric deterministic relationship among media, public opinion, and political agendas. Subsequent adaptations (e.g., Shoemaker 1989) recognize the influence of political structures, institutions, and readership on mainstream media practices. As noted in the preceding chapter, Soroka (2002) has to date offered the most nuanced, multi-faceted model for understanding how media, public opinion, and political agendas are derived in Canada. His model recognizes the inter-relationships among various forms of media, various groups within society, the changing structures and processes within government, and what he identifies as "real-world factors" (3), or occurring instances of an issue in the public sphere.

The agenda-setting model is difficult to apply in a predictive way, but it highlights the many points of influence where non-profit organizations can bring attention to issues within the public and political spheres. In its most basic form, the model suggests that those organizations that are successful in promoting their issues widely through the media will encourage public and policy debates on those issues. In its expanded form, it suggests that organizations can further influence this dynamic by capitalizing on emerging crises and issues to gain media attention and by communicating through various forums to increase and maintain their issues on the public and policy agendas.

Increased salience of an issue among media, publics, and policy makers does not ensure that the subsequent discussions are in alignment with an organization's mission or goals. Consequently, how an issue is framed (Entman 1993) or presented across these agendas is equally important to non-profit organizations. Research shows that news reports on social issues regularly present issues in an episodic frame, focusing on discrete events or individuals, providing little background or contextual information for understanding the circumstances surrounding the issue. As discussed in Chapter 10, this pattern of coverage presents an overall lack of attention to the systemic or structural causes of the issue or the need for more effective policies to resolve it (Iyengar 1990). At their most pervasive, the agenda-setting power of these media frames can result in distorted understandings of the issue among publics and legislative inaction to resolve the issue (Klodawsky 2004). It thus becomes even more important for non-profit organizations to actively pursue media exposure that will present their views and policy stances on specific issues.

Attracting media attention and influencing the media frame is challenging for many non-profit organizations. Todd Gitlin's (1980) seminal study of the Students for a Democratic Society and Robert Hackett's (1991) Canadian study of the 1980s peace movement offer insights into the difficulties faced by social movement organizations and special interest groups in raising and defining media issues. With little history or legitimacy, these resource-poor groups (McNair 2007) must create public spectacles or sensationalized events to be recognized by the media. What often result are media reports on the spectacle rather than on the issue, minimizing or even criminalizing the issues or views of the organization within the frame of coverage (Gitlin 1980). Subsequent research shows that these frames are not always consistent across the differing issues and mobilization stages of social movement organizations (Della Porta and Diani 1999), and efforts to be more strategic communication managers and actively pursue media coverage that educates the public on the cause and long-term solutions for the issue can counteract these frames or at least raise public debate surrounding the issue (Salzman 1998; Delicath and DeLuca 2003). As a result, many Canadian interest groups have adopted a sales-oriented marketing approach to their advocacy, using diverse methods to attract government and public attention to promote their organization and cause (Foster and Lemieux 2012).

Communicating their issue and perspectives through the media is more challenging for charity and service non-profit organizations. Research conducted in Canada and the United States found that non-profit organizations receive very little mainstream media coverage (Jacobs and Glass 2002; Greenberg and Walters 2004). Of the media attention that they do receive, non-profit organizations are more apt to have their deeds rather than their ideas reported on, as journalists often turn to the sector to provide localized, community, human interest stories and not as sources on hard news or issues stories (Deacon 1996, 1999). What result are patterns of coverage where "non-profit citizen organizations [are] not noted in either the cause, effect, or responsibility dimensions" of social issues (Kensicki 2004, 66). In the rare event that the views of non-profit organizations are featured, they are usually included as secondary sources, reacting to or confirming the comments of others.

This pattern of coverage has been attributed to both internal and external factors. Stretched budgets and a lack of staff, communication knowledge, and skills limit opportunities to actively pursue media coverage (Dimitrov 2008; Greenberg and Grosenick 2008; Foster and Lemieux 2012). The twenty-four-hour news cycle and understaffed newsrooms require immediate access to sources and information on the issue to help journalists meet these pressures (Manning 2001), and few non-profit organizations are skilled in these practices. Journalists are often uncertain about the credibility of the organization to speak about matters of policy (Deacon 1999), turning instead to more traditional government and corporate sources. To overcome these limitations, non-profit organizations have been encouraged to build media capacity within their organizations and to actively pursue media attention (Bonk, Griggs, and Tynes 1999; A. Davis 2003; Greenberg and Grosenick 2008; Dimitrov 2009). In addition to raising the profile of the issue, media attention can impart legitimacy on the organization, which can influence donations, volunteer support, program activity, and so on (Deacon 1999; Dimitrov 2008). Research shows that few organizations in Canada have developed this capacity or implemented basic practices to support media relations (Greenberg and Grosenick 2008). News reports are few and far between for most organizations, though the larger, better-resourced ones with in-house communication capacity are able to generate more coverage (Jacobs and Glas 2002, 245).

New and social media offer opportunities for non-profit organizations to communicate with citizens and decision makers. They also can support organizations' media management efforts. The use of social media is growing among publics and politicians (Grant, Moon, and Grant 2010). Similarly, journalists have embraced social media to solicit story ideas and story sources (Waters, Tindall, and Morton 2010). Journalists also regularly turn to organizations' Internet sites for background research and information when developing stories (Yeon, Choi, and Kiousis 2005; Greenberg and MacAulay 2009).

Non-profit organizations have yet to effectively employ new and social media. D. Jamieson (2009, 26) claims that, with few exceptions, Canadian non-profits' web presence is "half-hearted, unsophisticated and largely in-effective." Most organizations have a website; however, the information on it is often obsolete, rarely updated in a crisis, and fails to provide information of interest to journalists. There are few opportunities for public discussion or feedback, or even links to the organization's social media outlets (Hart, Greenfield, and Johnston 2005; Kenix 2008; Greenberg and MacAulay 2009). Similar criticisms are directed at non-profits' activities on Facebook and Twitter. Non-profits regularly open Facebook accounts but then aban-don them, unable to support them, and thereby turning off potential sup-porters (Waters et al. 2009). Most non-profits use Twitter as "just another way to send out information such as that found in traditional newsletters [or a] media kit" (Lovejoy, Waters, and Saxton 2012, 317), as opposed to engaging in active dialogue between organizations and individuals exter-nal to the organization. Non-profit organizations are abandoning or under-utilizing social and new media forums and their potential to create and sustain media and public debate about their issues and to connect with ex-ternal audiences.

Although there has been extensive international research exploring the opportunities, challenges, and barriers for non-profit organizations attract-ing media coverage and communicating through new media forums, there has been little exploration of how non-profit organizations are integrating these practices into their regular communication activities and the value they place on them. One Canadian study (Greenberg and Grosenick 2008) surveyed non-profit agencies in Ottawa to identify communications cap-acity and activity within the sector; however, this study does not inquire about new and social media activity. More recently, Foster and Lemieux (2012) interviewed nineteen special interest groups in Quebec to examine political marketing tactics. Still, there remains a general lack of understand-ing of how non-profit organizations communicate during their regular course of business, the goals they have for this communication, the tactics they use, and the degree to which they are strategically integrating and pri-oritizing agenda-setting activities to accomplish advocacy goals.

Case Study

Method

An online survey was conducted in 2012 to understand Canadian non-profit organizations' communication goals, tactics, and outcomes. The sur-vey was developed and prepared in both English and French for response via

SurveyMonkey. Volunteer Canada, a national organization representing non-profit volunteer organizations in Canada, partnered in the research and distributed the invitation to participate to their members. Two email invitations bearing the subject line "Communications Survey" were sent out to a distribution list of over nineteen thousand people. The survey was held open for four months. Participants were assured that their answers would remain confidential and reported only in aggregate. The survey instrument consisted of twenty short questions in four sections. The first section asked organizations to identify primary, secondary, and unfulfilled goals for their communication practices. The second section asked organizations to identify the form and frequency of public communication practices used in the past year. The third section inquired about which tactics organizations felt were most effective for achieving their primary and secondary goals. Questions related to communication capacities and demographics were included in the fourth section. The questionnaire was unique to the study, though some questions were adapted from a 2008 communications capacity survey of Ottawa non-profit organizations (Greenberg and Grosenick 2008).

A response rate of 6.2 percent was achieved (n = 118), including 109 English-language surveys and 9 French-language surveys. This response rate does not allow for findings to be generalized to the non-profit sector as a whole. It is believed that the funding and capacity challenges common to smaller organizations impacted the response rate and skewed the sample toward organizations with increased resources and capacities. Still, there was good diversity in the sample, allowing for findings to provide a snapshot of non-profit activity across the sector. The organizations that responded represented a wide range of focus, including social services, grant making, community-based health, education and research, arts and culture, business and professional associations, development and housing, religion, and the environment. Of the responding organizations, 73.1 percent identified a primary mission of program delivery and service, 20.4 percent identified a primary mission of public education and advocacy, and 6.5 percent identified a primary mission of trade or industry promotion. This distribution closely parallels US statistics that estimate that 25 percent of all non-profit organizations have an advocacy or lobby mandate, whereas the other 75 percent are focused on service, charity, and hobbies (Jacobs and Glass 2002). Approximately 75 percent of the organizations were registered charities, and the majority were provincially or federally incorporated. A wide range of operating budgets were represented, with many under $500,000, but a few with more than $10 million. The survey respondent was most often the executive director or communications manager. Two-thirds of the organizations reported that they had an individual connected with the organization dedicated to the communications function, but only 38.5 percent reported that this function was full time, and only 20.7 percent reported that communications was the individual's only responsibility.

Findings

Although the survey sample cannot be generalized to the non-profit sector as a whole because of the low response rate, the findings offer important baseline insight into the communications activities of the Canadian non-profit sector and how organizations within it define and prioritize their advocacy role. These findings are an important first step in understanding the goals and motivations of the non-profit sector for their communication activities, how they communicate on a day-to-day basis, and whether these activities support their advocacy role in Canadian political communication.

Overall, the study affirmed that many Canadian non-profit organizations' communication efforts are plagued by structural and capacity issues that limit their opportunity to advocate. At the same time, it revealed that there is little strategic planning and coordination devoted to the communications function to maximize the political communication opportunities available to them. The key findings of the study were threefold. First, Canadian non-profit organizations prioritize communicating with citizens to promote their organization and activities. Issue advocacy, conversely, does not register as a priority. Second, Canadian non-profit organizations actively publish information in both online and hard-copy formats. They are not as active in proactively pursuing traditional media coverage of their organization or issue. Finally, Canadian non-profit organizations believe that unmediated, internally produced communication pieces directed to existing or limited audiences help them achieve their public communication goals.

As noted, the survey asked non-profit organizations to identify their primary, secondary, and unfulfilled goals for communications. Figure 11.1 shows that, among the respondents, public awareness of the organization and promotion was the greatest priority. Client and membership support and development was second most prominent, and drawing in new funding was also important. Public policy advocacy, political awareness, and raising the profile of the organization as an issue spokesperson were of much lesser concern.

Figure 11.1

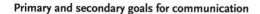

Primary and secondary goals for communication

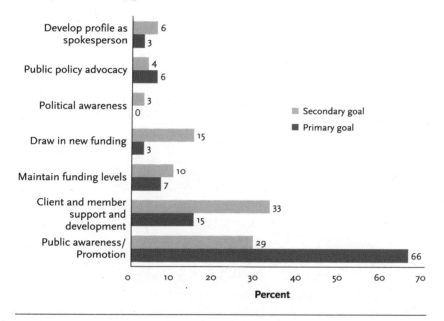

It is evident that non-profit organizations recognize the importance of publicly communicating about their organization. The survey did not ask what benefits these organizations felt would result from this increased awareness, but the need to promote the organization and its issue was accepted as an essential activity that organizations must engage in.

The focus on client service and delivery is not surprising given the large percentage of respondents with a primary mandate of service and program delivery. Project-directed funding and government service contracts often force these types of non-profit organizations to focus the majority of their administrative and communication efforts internally to service delivery activities (Dimitrov 2008). With limited staff and capacities, what often becomes the challenge is also directing effort to external communication efforts that will raise the public profile of the organization and issue and support the long-term viability of the organization (Deacon 1999).

The nominal focus on public policy advocacy reflects the structural and political environments that discourage advocacy for charity organizations specifically and for non-profit organizations generally. For those organizations that did report public policy advocacy as a priority, not all had a primary mandate of public education and advocacy; more than half (57 percent) were service delivery organizations, and 85 percent had registered charity status. This suggests that a small group of non-profits recognize the importance of advocacy, despite the many challenges and barriers limiting this activity.

Respondents were asked to identify the range and extent of their organizational communication activities over the past year. Consistent with recent studies (Foster and Lemieux 2012), non-profit organizations engage in diverse political communication activities. The most frequent activities reported were updating the organization's website, updating the organization's Facebook page, tweeting, and sending email updates and messages. The overrepresentation of organizations with higher levels of resources in the sample likely impacted these findings. These types of organizations would be more likely to engage in new and social media and to adopt new technologies. Website updates were regular occurrences (conducted at least bimonthly) for 67 percent of the organizations, with less than 9 percent of the organizations reporting that they did not engage in this activity at all. Conversely, there was a clear delineation between those organizations that actively engaged in Facebook and Twitter and those that did not. An equal number of organizations reported tweeting on a regular basis (42.1 percent) as those reporting not at all (42.1 percent). About half of the organizations regularly updated their Facebook page, whereas about one-third of the organizations reported no activity in this area. Although we cannot comment on the content of the websites, it is apparent that Canadian non-profit organizations are recognizing the importance of their web presence and seeking to maintain it, contradicting previous research that identifies non-profit websites as stagnant and obsolete (Kenix 2008; D. Jamieson 2009). The recent introduction of user-friendly webpage design platforms that no longer require non-profit organizations to outsource this function may help explain this trend. These findings also contradict previous research that claims few organizations are actively engaging in social media (Waters et al. 2009). This is likely a result of social media having larger audiences and being more prevalent in the mediascape at the time the survey was conducted than in the previous research. The increased audience and prevalence pressure more organizations to be active in these forums.

Hard-copy and face-to-face formats remained a key communications activity for the organizations in the study. Almost 50 percent continued to publish a newsletter on a monthly or bimonthly basis. Thirty percent printed and distributed program material at least once per month, and 35 percent printed organizational promotional pieces at least once per month. Forty percent regularly participated in events as an invited speaker, and another 21 percent organized public presentations at least once per month. These activities represent some of the more traditional non-digital, non-profit communication activities. These activities may be useful and beneficial for the organization, yet it is likely that many organizations undertake them with little strategic thought, because of their limited capacities. When faced with competing demands and little time for skill or professional development, part-time communication staffers are more likely to perpetuate existing activities than to explore new communications opportunities or formats.

Organizations in the study reported a moderate level of media management, though this activity was more reactive than proactive. Whereas almost half reported responding to media inquiries on a regular basis (more than six times per year), only 34 percent reported sending press releases on a regular basis, and even fewer reported organizing press conferences or media events. These findings suggest that there is a desire and willingness to support media reports about their organization; however, organizations are not actively facilitating and promoting them. Few organizations had a goal of developing or maintaining a profile as an issue spokespeople or commentators. Only 20.7 percent regularly pitched stories to journalists, and only 22.5 percent offered media training for their spokespeople. The data support previous research that identifies limited organizational capacities (Greenberg and Walters 2004) as barriers to proactive strategic communication. However, the passive approach to media management limits the opportunities available to organizations to raise the salience of their issues and make their perspectives known publicly and, by extension, among policy makers, as the agenda-setting model suggests. For those organizations in the study that did report an active media strategy, the survey data did not allow for an in-depth analysis of those organizations' funding and capacity structures. This would be important for future research.

Communication with policy makers was also virtually silent on other fronts. Some organizations reported periodically sending material to government officials, but few organizations initiated meetings with government staff and fewer still initiated meetings with elected representatives. More-over, few organizations participated in an advocacy coalition. Alliances and coalitions are efficient and cost-effective ways for organizations to address community needs, pool resources, and more effectively achieve advocacy mandates (Cruz 2001; Gormley and Cymrot 2006). More importantly, co-alitions can provide a sense of security to non-profits concerned about their funding or the political implications of lobbying (Cruz 2001). This tactic has yet to achieve significant uptake among Canadian non-profit organizations.

Non-profit organizations were invited to identify which communica-tion tactics they felt most contributed to their primary and secondary goals. For those organizations that listed public awareness and promotion of the organization as their primary goal, the communications tactics iden-tified as most important to achieve this goal were producing and distribut-ing a newsletter, updating the organization's website, publishing material on the program or organization, and emailing updates to members. Similar activities were also identified as most effective for achieving the goal of client and membership support and development; these included emailing up-dates to members and updating the organization's website and Facebook page (see Figure 11.2). The common responses to achieve both external and internal goals underscore the findings that non-profit organizations are not strategically managing their communication programs and have unachiev-able expectations for many of their communication efforts.

Producing newsletters and program information, emailing existing members, and updating the organization's website are effective tactics to communicate with and maintain support among existing stakeholders; how-ever, they are limited in creating public awareness and promoting the organ-ization to wider audiences. Existing stakeholders and supporters are the primary audience for newsletters and client emails. Published organization and program information can reach individuals beyond the organization; however, the hard-copy format, which requires time, cost, and opportunity to distribute, limits its reach. Similarly, as a pull medium, websites most often communicate with existing audiences (D. Jamieson 2009). Websites can be an important source of information for citizens and journalists in times of emerging crisis and issue salience; however, non-profit organiza-tions must proactively pursue media coverage of their organization to gain this traffic. Moreover, the unmediated nature of communication directly af-fects the legitimacy of the organization (Towner and Dulio 2011). Non-profit organizations are not targeted or strategic in their communication

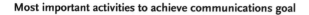

FIGURE 11.2

Most important activities to achieve communications goal

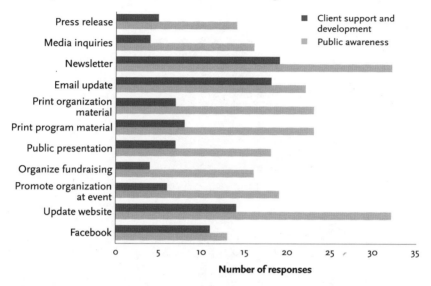

Number of responses

practices, aligning tactics with desired audiences, anticipated goals, and outcomes (Spitfire Strategies n.d). As Foster and Lemieux (2012, 167) found, "[audience] targeting does not seem to be a strategic priority."

Political Communication in Canada

Non-profit organizations are important political communicators in Canada. They often speak for the most vulnerable citizens and offer grassroots, front-line views about social issues and practices that can increase public understanding of issues and contribute to effective public policy development. Yet Canada's non-profit sector has difficulty communicating these views. Many non-profit organizations, most prominently the smaller and less-resourced service-based organizations, employ a shotgun approach to their communication activities. They spend little time targeting and planning their efforts or coordinating messages across the various communication forums employed. This does little to ensure the messages reach and resonate with large public audiences and legislative decision makers. More research is required to correlate the organizational characteristics and practices that complement the development of an effective, strategic communications function for non-profit organizations within the sector to help them achieve their advocacy goals.

There exists real concern and evidence to suggest that government structures and processes are actively excluding non-profit organizations from the policy process. The agenda-setting model posits that non-profit organizations could influence social and political change by increasing the salience of their issue and how it is understood in public opinion and through mainstream, and possibly new and social, media. However, be it because of capacity issues or a lack of knowledge of the agenda-setting function of the public media, for many organizations, these opportunities are being missed.

A. Davis (2003) argues that mainstream news journalists are turning away from traditional news sources such as corporations and governments, offering unprecedented opportunities to non-profit organizations to raise and define issues in the news. Yet attracting mainstream media coverage is not a priority for many non-profit organizations, especially the smaller, less-resourced groups. There is little evidence that, when contacted by journalists, they capitalize on this attention by providing professional messaging and information in a format that supports media practices and processes and helps frame the issue in a way that supports their mandates. This perpetuates the scarcity of coverage of the non-profit sector, journalists' uncertainty about non-profits' credibility to speak on issues, and public discourses that support the status quo.

Websites play a critical role in non-profit communication with stakeholders and can support organizations' efforts to inform publics and journalists about issues. Evidence suggests that non-profit organizations are more actively tending to their web presence; however, more research is required to analyze the content of these websites. Websites can help organizations maintain an active public profile and provide timely content that educates individuals seeking out information and assists their media relations efforts (Yeon, Choi, and Kiousis 2005). This information must be formatted to be easily accessible to multiple audiences. If it isn't, opportunities to contribute to the public debate, or at least inform external publics seeking out the information on the Internet, are missed.

Social media is becoming more common in the non-profit communications tool box, though only about half of the organizations in the survey are regularly engaging with it. Effective social media management requires constant and sustained activity and use, which challenges the ability of many organizations to actively participate. Research suggests, however, that opportunities exist for non-profit organizations to utilize these forums to expand and build relationships and networks of supporters; to provide regular information about the organization to supporters, journalists, and policy makers; and to potentially engage in debate on their issues. They are also becoming more important sources of information for publics and decision makers and thus cannot be ignored. For those organizations that have not yet developed a social media presence, these opportunities are being missed.

With web-based platforms, information flows are more immediate and accelerated (McNair 2009). It is from this information that citizens and legislators identify priorities and how best to address the issues that they prioritize. To compete for the attention of the public and decision makers, non-profit organizations must become more active in the expanded mediascape to be regarded as a credible source by mainstream media, and to produce and deliver information to citizens and legislators to inform their decisions. This requires an understanding of how to construct messages to present their views on an issue and to gain the attention of desired audiences, and the knowledge and skills to identify and manage the available communication tactics to deliver these messages. Increased activity and profile, however, can be a double-edged sword. They offer increased opportunities for unintended or negative communication effects and open the organization up to challenges from dissenters and opponents. Hence, non-profit organizations must not only attract and produce coordinated communication pieces but must also be equipped to monitor and manage public deliberation on their issue and organization.

This is a tall order for many groups. Non-profits continue to be plagued by funding and staffing issues that limit their ability to dedicate skilled resources to communication planning and management. Project-specific funding and government contracts concentrate efforts on internal audiences and short-term issues. Press releases and messages are generated based on past practices or the limited knowledge of the part-time staff or volunteer tasked with communication. Non-profit strategic communication can no longer be viewed as an administrative luxury. To achieve their political and social mandates, non-profit organizations of all sizes must find ways to centralize public communication as an essential function and more effectively facilitate their communication activities.

12

Blogging, Partisanship, and Political Participation in Canada

Thierry Giasson, Harold Jansen, and Royce Koop

The decline of real-world political participation in Canada is by now an old story. Turnout has declined noticeably, particularly as the electoral abstinence of those born in the 1960s and 1970s has manifested itself in the overall turnout rate (Blais and Dobrzynska 2003). Low rates of other types of political participation accompany low turnout rates. This is not surprising in the case of party membership and activism, as Canadian parties have not generally been organizations that have empowered their members (Carty and Cross 2010). Although the parties showed some inclination to engage members in the 1990s (e.g., Young and Cross 2002), there is little evidence that this democratic impulse has continued to the present day, given that their use of social media tends to be foremost for broadcasting than for engaging (see Chapter 6).

In contrast, online political participation has increased in recent years as Canadians have used the opportunities presented by the Internet to learn about politics, debate political issues, and post their opinions online. One of the earliest and most popular forms of online political participation in Canada is blogging. Blogs became more widespread following the introduction of free blogging software from 2000 onward (Koop and Jansen 2009, 158), and their use grew exponentially in the United States (Rainie 2005) as well as in Canada during the mid-2000s. In 2010, a study reported that 33 percent of Quebec adults had read at least one blog and that 16 percent had kept a blog or posted comments on a blog in the past year (CEFIO 2010). Moreover, the prevalence of blogging activities had been increasing in the province over the last three years. This represents a substantial participation rate, even though other more passive forms of social media use seem more popular among Canadians. Fifty percent of Canadians surveyed in 2011 said that they had social networking profiles; of them, 86 percent maintained Facebook profiles and 19 percent had Twitter accounts (Ipsos 2011).

The Canadian political blogosphere is partisan in nature, with communities of bloggers united under groupings of blogs called "blogrolls" that are associated to each dominant political force in the country (Elmer et al. 2009; Koop and Jansen 2009). In 2004, activists developed the Blogging Tories blogroll community, which brought together Conservative bloggers in order to open an online front against the then ruling Liberal Party, as well as offer an alternative source of perspectives from those of the mainstream Canadian media (Koop and Jansen 2009, 159). Liberal and NDP activists followed in 2005, with the development of the Liblogs and Blogging Dipper blogroll communities. In Canada, political bloggers tend to be defined by partisanship rather than ideology, rendering blogging an important online extension of partisan political participation.

This raises the question of the relationship between real-world and online political participation. Several studies address the topic indirectly, but none addresses the question of how real-world participation conditions the online behaviours and perceptions of political bloggers (and vice versa). Several studies have focused on the ease with which citizens can participate in online political activities, and have as a result speculated that lower costs lead to enhanced participation in such activities. Following Di Gennaro and Dutton (2006) and Xenos and Moy (2007), Borge and Cardenal (2010) found that Internet users require less motivation to participate in online political activities. Indeed, experienced web users are likely to get involved in these activities, independent of the other motivations typically thought to be required for political participation. Although not explored, the consequences of Borge and Cardenal's findings appear to be that online participation – including partisan blogging – is a low-cost alternative to real-world participation such as participating in election campaigns, suggesting that online participation replaces real-world participation.

The web provides opportunities for citizens to replace high-cost real-world participation with low-cost online participation, yet online resources are in fact related to participation in real-world political activities (e.g., Krueger 2002). Tolbert and McNeal (2003), for example, demonstrate that Americans with access to the Internet and online news were significantly more likely to vote than were respondents without such access, even when controlling for a range of variables traditionally associated with the decision of whether to vote. Most importantly, Internet access is also related to increased participation, including discussing politics with others, attending rallies, working for campaigns, and contributing money to candidates, parties, or interest groups (Tolbert and McNeal 2003, 182). Other studies have found relationships between Internet access and real-world participation, including contributing financially to campaigns (Bimber 2001), attending rallies and signing petitions (Weber, Loumakis, and Bergman 2003), and having contact with elected officials (Bimber 1999; Stanley and Weare 2004). Use of the Internet is also related to civic engagement and contentment (Shah, Kwak, and Holbert 2001). These studies focus on results from surveys of adults, yet Quintelier and Vissers (2008) demonstrate that Internet use is also related to real-world participation among youth.

There appears to be a growing consensus in the literature that real-world and online participation are related, though the causality underlying this relationship (if any) is not always clear. This chapter explores the relationship between online and offline political participation in Canada (with particular emphasis on blogging) and clarifies both common and divergent motivations to participation on the web and in the "real world." Accordingly, we advance the scholarly understanding of the relationship between real-world and online political participation by first comparing the behaviours and perceptions of 135 Canadian political bloggers who engage in different types of real-world political participation, and then by identifying their motivations and intentions to blog about politics.

Case Study

Method

This chapter's dataset was collected through a secure online survey available on the Groupe de recherche en communication politique's (GRCP's) website from February 1 to 28, 2011. The online questionnaire featured seventy-six questions distributed in eight sections addressing specific themes such as bloggers' socio-demographic characteristics, their political allegiances, their blogs' content, and their blogging practices, as well as their communication objectives and intentions. For instance, respondents were asked why they blogged about politics, how long had they been blogging, where they got their inspiration from, what forms of political actions they engaged in, who they write for, and what type if any influence they felt they had on Canadian politics. All the participants in the study were Canadian adults, living in Canada, with the right to vote and who were contributing weekly to at least one political blog.

The strategy developed to build the sample was twofold. First, a conventional reasoned choice approach was employed to select an initial sample of the 170 active political bloggers from all regions of Canada and all partisan affiliations (49 associated with Liblogs, 45 with Blogging Tories, 25 with Blogging Dipper, and 51 from Quebec's French-language blogosphere). They were identified as the most active blogs in the pool of partisan blogs identified by Ryerson University's Infoscape Research Lab. Email invitations to participate in the survey that included a hyperlink to the online questionnaire were sent to these selected bloggers both in French and English. Weekly reminders were forwarded during the survey period to those who still had not completed the survey.

Second, a snowball approach was employed to circulate calls for participants within the Canadian blogosphere, through online and offline networking. Recent studies have demonstrated that the structure of online communication favours the viral circulation of mixed-media information (Viégas 2006; Jankowski and Van Selm 2008, 6; Sweetser and Lariscy 2008, 179). Our snowball technique, quite similar to online viral diffusion, was accomplished through three distinct communication channels. First, those bloggers contacted via the initial email invitation were asked to forward the survey's web link to three other Canadian political bloggers they knew through their personal online or offline network or from their daily online media consumption. They were also invited to publicize the study by posting a web link on their blog redirecting their readers to the online questionnaire: a method that has been employed previously by other scholars (Johnson and Kaye 2004). Second, the call for participants was circulated on two Twitter feeds dedicated to Canadian (#cdnpoli) and Quebec (#polQC) politics. Third, an email was sent to thirty Canadian political reporters and commentators, from French- and English-language news organizations, who publish widely read blogs to ask them to publicize the study's call for participants. Our interest resides in citizen political bloggers and their political participation. Consequently, all email invitations and the circulated open call for participants explicitly stated the exclusion of professional journalists and pundits who blog about politics for their respective media outlets from the study.

The viral dissemination of survey invitations enables a wider circulation within the political blogosphere, though some scholars have pointed out that it can generate selection effects with detrimental impacts on the validity of the results (Jankowski and Van Selm 2008, 6). For instance, it can motivate individuals with a political or research agenda to fill out the questionnaire, therefore tainting the representativeness and, more importantly, the validity of the data collected (ibid.). Nevertheless, Babbie (1998) believes that viral non-probabilistic sampling strategies, which are primarily used in qualitative investigations similar to the one carried out in this study, are appropriate for identifying special populations who are difficult to locate by other means.

A total of 148 bloggers ultimately filled out the online questionnaire during the four-week recruitment period. More specifically, 87 of the 170 most active political bloggers contacted in the first sampling round answered the survey. In addition, 61 bloggers reached during the viral dissemination phase, through social networks and mainstream media coverage of the study, completed the survey.[1] Respondents were from all over Canada, but a large subsample of them resided in the three most populous provinces in the country: Ontario (40 percent), Quebec (37 percent), and British Columbia (8 percent).

Findings

Our survey investigated the demographics of bloggers, their blogging practice, their political participation, and their motivations to blog. We relied here on their self-reported behaviour, not on an analysis of their actual behaviour through a content analysis of their blogs. To better investigate how partisanship may affect the practice and perceptions of political bloggers in Canada, we first develop a typology of partisan commitment levels. We then outline the socio-political portrait of those bloggers (demographic characteristics, blogging practice, audience, style, and perception of impact). Through all of this, we are interested in the extent to which the respondents' partisan commitment influences how they blogged about politics. Finally, we look at what motivates Canadians to blog about politics and which specific goals they wish to achieve by blogging the political.

Using a battery of survey questions we polled respondents about partisan commitment and involvement. We then divided these bloggers into three categories. First, we identified a group of *non-partisan bloggers*. These respondents indicated that they did not identify with any political party, either federally or provincially ($n = 22$; 16 percent). The second category consisted of those bloggers who indicated identification with either a provincial or federal political party (or both) (n = 47; 34.8 percent). We refer to these people as *identifiers*. The third category included bloggers who both identified with a particular political party and were party members (n = 66; 48.9 percent). We refer to these people as *members*.[2] These three categories represent a scale of increasing partisan involvement and commitment – from non-partisans to identifiers to members.[3] Overall, these bloggers show a significant level of partisan involvement. Nearly half of our respondents reported being party members, much higher than the traditional estimates of 1 to 2 percent of the Canadian population (Cross 2004). Only about a sixth of our respondents did not indicate a partisan identification, about half the proportion in the Canadian population (Gidengil et al. 2012, 61).

Although our primary interest here is whether bloggers with different levels of real-world partisan involvement approach their blogging differently, we also examine whether this distinction in partisan involvement among bloggers is related to other demographic characteristics. Table 12.1 reports the gender of our respondents, broken down into the three partisan categories. We first note that the vast majority of Canadian political bloggers are men. The proportion of women found in each category decreases as the level of partisan commitment increases. Put another way, the few women in our sample of bloggers were essentially evenly divided among the three categories; it was the men who largely account for the variation in numbers in levels of partisan commitment. These conclusions are consistent with the findings of Cross and Young's survey of party members, which found that party members were disproportionately male (Cross 2004; Cross and Young 2004), and with previous studies of other national blogospheres indicating that the vast majority of bloggers are male, with the political blogspace being even less feminized (Giasson, Raynauld, and Darisse 2011; McKenna and Pole 2004; Pedersen and Macafee 2007). Even though the number of women

TABLE 12.1

Gender of bloggers by partisan commitment

Partisan commitment	Non-partisans (%)	Party identifiers (%)	Party members (%)	Total (%)
Female	28	14	8	13
Male	72	86	92	87

NOTE: n = 127. Rounded figures.

bloggers within the general blogosphere has been on the rise in recent years, women political bloggers remain the exception. The average age of our sample of respondents is forty-one. Unlike Cross and Young (2004), who found that party members were significantly older than the Canadian population, our data indicate that bloggers who are party members are more likely to be younger than those bloggers with lower levels of partisan commitment.

When it comes to socio-economic status, we found that different levels of partisan commitment were associated with different levels of educational achievement and income. Regardless of partisan commitment, our respondents form a well-educated group. Over 70 percent of bloggers in all three categories have at least one university degree, and fewer than 10 percent of bloggers in each category have a high school diploma or less. Overall, however, the non-partisan bloggers were better educated than either the identifiers or the party members. Nearly 60 percent of the non-partisans reported having graduate or professional degrees, compared with about a third of the identifiers and the partisans. The partisan identifiers, however, were distinctive in their income, as they were less likely to be affluent; only 10 percent reported family incomes above $100,000, compared with 40 percent for the other two groups. Consistent with the findings on education and income, the bloggers in our sample were employed mostly in skilled and technical fields. The most common occupational fields were education (13.1 percent), communications and marketing (11.7 percent), and information technology (10.3 percent). Only two of our bloggers indicated working in "the media," and none self-reported that he or she was a journalist. Our examination of occupation revealed no particular differences between the three categories.

Our sample of respondents shows a larger proportion of left-leaning and moderate bloggers. Close to 30 percent of surveyed bloggers self-identified as right-leaning (e.g., conservative, Christian democrats, libertarian), whereas 18.9 percent identified themselves as centrists (liberals, moderates), and 41.9 percent leaned left (progressives, socialists, social democrats, communists). Larger proportions of members and identifiers than non-partisans could more easily state an ideological stand. Party members were slightly more conservative than the other two categories of bloggers, identifiers more progressive, and non-partisan more centrist. These results constitute a very sharp distinction from the ideological structure of the political blogosphere in the United States. Bowers and Stoller (2005, 7) argue, based on several analyses of the US political blogosphere, that the conservative blogging community "was between two and three times as large as the progressive [or liberal] blogosphere" in 2003. Our findings are more closely aligned to those highlighted in previous analyses on Quebec political bloggers (Giasson, Raynauld, and Darisse 2009, 2011). Our respondents' ideological positioning seems to indicate that Canada's political blogosphere may be more progressive and liberal than its American counterpart.

The demographic portrait that our survey allows us to cast of the Canadian blogosphere is quite similar to profiles produced in other national contexts. The differences between levels of partisanship proved to be relatively minor. We expect, however, to find greater distinctions in the approach to blogging taken by each type of blogger. To this end, we asked several questions in order to understand how bloggers approach their writing.

First, we asked whether bloggers post under their real names, nicknames associated with their real names, or pseudonyms. A majority of respondents (61.2 percent) in all three categories post under their real names. The most significant difference is that the identifiers were the least likely to post under their real names and the most likely to post under a nickname. There were no large differences between the categories in many other measures of blogging behaviour. About 70 percent of the respondents in all three categories had been blogging about politics for three or more years. As for blogging frequency, bloggers in the identifier category posted slightly less frequently than bloggers in the two other categories.[4]

Next, we asked Canadian political bloggers what proportion of their blog posts was political. For all three categories, the bloggers' primary focus is on politics. On average, 86 percent of the members' posts were on politics, as were 82 percent of the identifiers' posts. The non-partisans were less likely to post about politics, with 73 percent of the blog posts being political in nature. Other topics of interest mentioned by respondents were the Internet and information technology (35 percent), the arts (29 percent), non-political news (25 percent), and business news (21 percent). Finally, just under a quarter of surveyed bloggers said that they regularly write about their personal lives.

It is when we move to items tapping into blog style and audience that we observe more significant differences between the three partisan groups of bloggers. We asked respondents who they blogged for. As shown in Table 12.2, the major difference is between the non-partisans and the identifiers

TABLE 12.2

Political bloggers' intended audience

	Partisanship (%)		
	Non-partisans	Party identifiers	Party members
Yourself	46	62	67
Friends	18	28	30
People with similar political views	36	60	64
Journalists and traditional media	36	34	44
Other bloggers	36	53	49
General audience	54	60	62
(n)	(22)	(47)	(66)

NOTES: Rounded figures. Columns may add up to more than 100% because multiple responses were allowed.

and members. Identifiers and members were more likely than non-partisans to write posts for friends, people with similar political views, and other bloggers. There appears to be a "preaching to the converted" element to those bloggers, with a partisan leaning that is absent among non-partisans, furthering the hypothesis associating political blogospheres to ideological "echo chambers" (Wallsten 2005). The partisan bloggers are also more likely to report writing for themselves than were non-partisans, suggesting that external motivations might be slightly more important for non-partisan bloggers than for others.

We also asked bloggers to indicate who they thought read their blogs. As Table 12.2 shows, a majority of bloggers in all categories wrote for a general audience, though non-partisan bloggers were more likely to report an audience of "regular Internet users." Conversely, the two categories of partisan bloggers were more likely to perceive that like-minded citizens were reading their blogs. Again, there are a couple of ways in which party members differ from the other bloggers: they are less likely to perceive that other (non-political) bloggers are reading their posts, and they are more likely to perceive that opposing political activists are reading their blogs.

Given that our respondents vary in their intended and perceived blogging audiences, we expect that the way they blog should vary accordingly. We asked them to indicate where they got their inspirations for blog posts. Table 12.3 shows that nearly all bloggers who answered the survey derive

TABLE 12.3

Inspiration for blog posts

	Partisanship (%)			
	Non-partisans	Party identifiers	Party members	Average
Political coverage in traditional media	90	94	96	93
Current political events	73	72	86	77
Other blogs	64	66	83	71
Communication from parties	27	40	65	44
Contacts in political organizations	32	40	62	45
Entourage	36	36	42	38
Comments from the Internet	46	47	44	46
(n)	(22)	(47)	(66)	(135)

NOTES: Rounded figures. Columns may add up to more than 100% because multiple responses were allowed.

significant inspiration from political coverage in the traditional media and from current events. Blogging, like traditional news coverage, is driven by the events of the day. In addition, bloggers from our sample, despite their frequent derision of the mainstream media, derive substantial inspiration from the traditional media in writing their blogs. These bloggers are unique, however, in the inspiration they find in other blogs while writing their posts. As suggested by Scott (2007, 41-45) in the American context, bloggers frequently assume functions of surveillance and correlation (editorializing and commenting) with conventional news media. They are significant news junkies, and they relay in their posts numerous political news items coloured with their own interpretation of the strategies of news organizations or of political actors covered. This indicates that they pay attention to mainstream media, but that the attention of some bloggers to a particular issue or event will have a snowball effect, with other bloggers soon taking inspiration from those posts and focusing their writing on these topics.

Data show that as the level of partisan commitment increases, the importance of political communication from parties as sources of inspiration also increases. In addition, party members in the sample are much more

TABLE 12.4

Dominant tone in blog posts

	Partisanship (%)			
	Non-partisans	Party identifiers	Party members	Average
Partisan stance on issues	9	16	26	17
Attack and critique of opponents	23	42	27	31
Promotion and defence of allies	5	13	20	13
Neutral analysis	36	11	14	20
Other	27	18	14	19
Total (%)	100	100	100	100
(n)	(22)	(45)	(66)	(133)

NOTE: Rounded figures.

likely to report that their personal contacts in political parties are a source of fodder for blog posts. Given that the member category is defined by an involvement in a political organization, this is hardly surprising.

The most marked differences between our categories, however, have to do with the tone of blogging. Table 12.4 indicates that these blogs take a surprising variety of stances. Overall, the most common tone is "attack and critique of opponents," which is consistent with perceptions that political blogging is partisan and that the Internet is generally a negative space (see Jansen and Koop 2005; Koop and Jansen 2009). Despite that, some clear differences emerge. For instance, non-partisan bloggers in the sample are most likely to present what they consider to be neutral analyses of political events, and the likelihood of presenting a partisan stance on issues increases along with the degree of partisan commitment. However, members and identifiers differ in their approaches. Although attacking opponents is the most frequently reported tone for both members and identifiers, the identifiers are particularly likely to emphasize attacking and critiquing opponents – the ratio of those with a dominant negative tone to those with a positive tone is more than 3:1. Members, on the other hand, are more likely to be positive in their blogging: the ratio of negative to positive is 1.5:1.

Given the literature on the relationship between online and offline political participation, we also investigated the extent to which bloggers in our

TABLE 12.5

Forms of political participation

	Partisanship (%)			
	Non-partisans	Party identifiers	Party members	Average
Voted in the last federal election	68	91	94	84
Voted in the last provincial election	68	88	91	82
Traditional offline activity	14	35	94	48
Non-traditional offline activity	59	68	88	72
Traditional online activity	0	11	50	20
Non-traditional online activity	18	38	47	34
(n)	(22)	(47)	(66)	(135)

NOTES: Rounded figures. Columns may add up to more than 100% because multiple responses were allowed.

three categories engaged in other political activities. Table 12.5 summarizes the findings. Overall, the surveyed blogging population is quite active in other forms of political participation besides their blogging. The most important of these activities is voting in both national and provincial elections. However, bloggers in our sample were also likely to participate in what we term "non-traditional offline activism" and were more likely to do so than engage in more traditional forms of political participation.[5] This suggests that blogging is supplementing rather than replacing activism among this politically engaged group of Canadians.

Regardless of partisanship level, the surveyed bloggers voted in much larger proportions than do other Canadians, and they take part in a wide array of traditional and non-traditional forms of political engagement. They are "hyperactive" political citizens. However, non-partisans and identifiers are less involved in activities associated with political parties, such as online and real-world campaigning or taking part in partisan rallies and meetings.

Nonetheless, the level of political activism generally increases as the degree of partisan commitment increases. Simply put, party members are more active than identifiers, who are more active than non-partisans. For items that measure involvement in partisan activities, such as donations to

Table 12.6

Bloggers' motivations

	Partisanship (%)			
	Non-partisans	Party identifiers	Party members	Average
Take part in political debate	14	11	15	13
Voice opinions	43	74	60	59
Improve civic skills	0	13	8	7
Circulate information	10	11	13	11
Complement mainstream media	19	17	20	19
Contribute to democracy	19	4	6	10
Fight opposing ideas	5	9	10	8
Stimulate dialogue, reactions	14	9	3	9
Change things	24	13	19	19
Other motivation	29	11	15	18
(n)	(22)	(47)	(66)	(135)

NOTES: Rounded figures. Columns may add up to more than 100% because multiple responses were allowed.

parties or campaigning, this is not surprising. Nevertheless, it appears that even non-partisan activities are more common for party members and identifiers than for non-partisan bloggers.

What motivates Canadians to start writing political analyses on blogs? Respondents indicated that reasons to blog are deeply personal. Regardless of partisanship, the vast majority (61.4 percent) said that they blog to voice publicly their concerns and opinions. As Table 12.6 shows, 7 percent of respondents also described blogging as a way to develop civic skills, through regular reading, writing, reasoning; and news-gathering routines. Blogging on politics makes them better citizens.

Other surveyed bloggers expressed more instrumental motivations to justify their presence in Canada's political blogosphere. As previous researchers have found, 18.5 percent of surveyed bloggers wish to offer a complement to the political coverage of traditional news media. Many expressed strong and negative opinions about what they consider to be biased, unbalanced political news produced by the Canadian media and think of their blogs as alternative sources to mainstream news. Other bloggers said that their contributions were a way to take part in public debate (13.3 percent), to circulate political information they deemed important for society (11 percent), or because they wish to change society as a whole by influencing other citizens (19 percent).

When looking at the relationship between partisanship level and motivations to blog, we found differences between bloggers in our sample. First, larger shares of non-partisans than identifiers and members highlighted the altruistic motivations to "contribute to democracy" (19 percent), "stimulate dialogue and generate reactions" (14.3 percent), and "change things" politically (23.8 percent). Second, larger proportions of members (10.4 percent) said they blog to counteract opposing positions and issues, a very partisan motivation. Finally, even though this is the dominant motivation for all bloggers to post contributions online, a much larger share of identifiers than members and non-partisans indicated the need to voice opinions as their main motivation to blog. The data depict an active political community, ready for debate and public expression. Furthermore, the data show that this community is made up of individuals wishing to become more eloquent, more knowledgeable, and more reflective citizens.

The survey also investigated the communication intentions of Canadian political bloggers through an open-ended question: What type of communication do you wish to generate by posting contributions online? Respondents' stated communication strategies reflect their motivations to blog, with just over half of them wanting first and foremost to circulate information (Table 12.7). The surveyed Canadian political bloggers also stated other intentions associated with their political contributions online. Persuading others, debating, and exchanging ideas, as well as voicing opinions, are all mentioned as key communication goals by respondents. Oddly, only 7 percent of political bloggers stated partisan activism as one of their central communication objectives. This result differs from the significant proportions of respondents who stated using some form of partisan tone in their posts (64.4 percent) and who were volunteering actively in political parties or grassroots organizations (45.8 percent and 41.2 percent). Larger proportions of non-partisans than identifiers and members indicated that they wish to offer new perspectives on politics. Conversely, significantly more members than identifiers or non-partisans wished to engage in activism through their blogs or use them to persuade their audience. Finally,

Table 12.7

Bloggers' intentions and goals

	Partisanship (%)			
	Non-partisans	Party identifiers	Party members	Average
Information	62	52	45	53
Debate	19	37	28	28
Activism	0	9	12	7
Persuasion	24	33	45	34
Voice opinions	14	37	30	27
Offer new perspectives	14	7	8	10
(n)	(22)	(47)	(66)	(135)

NOTES: Rounded figures. Columns may add up to more than 100% because multiple responses were allowed.

more identifiers than bloggers with other levels of partisanship said their blogs are tools destined for political debate and voicing opinions.

Political Communication in Canada

Most research on blogging has focused on content analysis of blogs or on the consequences of online behaviour for real-world behaviour. Our analysis takes a different approach by investigating how real-world political commitments affect bloggers' approaches to their blogs. We have emphasized one possible determinant of this online behaviour: the level of partisan commitment. We found in our sample that bloggers' political affiliations help explain their blogging behaviour, styles, and motivations.

We identified three kinds of bloggers. Non-partisan bloggers were respondents who did not indicate an affiliation with any federal or provincial political party. Compared with the surveyed bloggers in the other categories, non-partisan bloggers tend to be older, more highly educated, and women. They express in higher proportions altruistic and idealistic motivations to blog, such as a will to change politics, contribute to democracy, or stimulate dialogue. Their blogs are aimed at a variety of audiences, but they are more likely to be read by a general public. These bloggers draw their inspiration for their messages from traditional media and current events. They are more likely to take a non-partisan tone in their blogging, emphasizing analysis of events, issues, and positions, rather than engaging in partisan to-and-fro. This appears to extend to their real-world political involvement as well. These respondents are most likely to see their blogging as having an impact on debate and public opinion.

The partisan identifiers in our sample, on the other hand, tend to have lower incomes than the other groups. They blog less frequently and spend less time on their blogs. They blog to voice their opinions and to improve civic skills. They wish to inform and debate with other bloggers. Unlike the non-partisans, they are most likely to blog for themselves, for friends, and for political allies, and they perceive their audience to be primarily the like-minded. Like non-partisans, they draw inspirations for their blog posts from the media and current events, but they are also influenced by communication from party sources. Their style is far more partisan, and these respondents are most likely to take a negative tone of attacking political opponents.

The party members make up the largest category of political bloggers in our sample. They tend to be younger and are most likely to be male. They aim their blogs more frequently at journalists and the traditional media, though they are skeptical of the impact they have on that audience. Their style is somewhat distinctive from the other two groups we studied in that they are the more likely to be influenced by communications with parties and to derive inspiration for blog posts from their contacts within parties. Members in our sample are more likely to blog for persuasive purposes, to consider blogging as a form of activism, and to perceive their role in partisan terms, though they are more likely to approach blogging as a positive activity of promoting a particular point of view. They also are the group that most values interaction with their audience in the form of comments.

This study reveals that online political participation through blogging does not replace offline real-world forms of engagement. It represents a supplementary type of participation. This research profiles bloggers who are very politically active, informed, and critical individuals. Indeed, they can be defined as politically hyperactive, taking part in numerous traditional and novel forms of political and civic engagement, ranging from voting and party membership, to public demonstration and boycotts, to web activism. They are more politically engaged than the general population. They are effectively "hypercitizens." Their motivations and intentions to blogs are diverse, but all seem to be grounded in a will to engage in political debates, to express opinions, to share information, and to improve such civic skills as gathering information about politics or developing strong arguments in support of issues. The surveyed Canadian bloggers do not engage in this activity lightly. They are profoundly dedicated to it, investing personal time, resources, and energy to post contributions they feel will make a difference to them, to other members of the blogosphere, and to Canadian society as a whole. Their endeavour is inherently political and democratic, rarely tainted by cynicism. Canadian bloggers who took part in this study are critical and skeptical, yet still very trusting of Canada's political system and democratic institutions, as their high level of electoral turnout attests.

The research helps us better understand how the new context of political communication in Canada – characterized by faster circulation of information, open access to communication technologies, and the central role strategy plays in message development – affects the way citizens relate to the political system and allows them to become active producers of political contents. Although many studies have examined different facets of the political blogspace in the United States, the potential influence of Canadian political weblogs on online, as well as offline political information production and consumption patterns, electoral campaigning, and policy-making processes, have been evaluated by only a handful of scholars (Koop and Jansen 2006, 2009; Chu 2007; Small 2008a; Giasson, Raynauld, and Darisse 2011). Much remains to be explained about political blogs and their authors in Canada.

NOTES

1   Although the total number of 148 respondents is smaller than other studies of general populations of bloggers – such as Kullin's (2008) 700 respondents or the 631 respondents in Braaten's (2005) study – the few studies that have built on interviews and surveys of political bloggers have been based on similar or smaller samples (Giasson, Raynauld, and Darisse 2011, $n = 56$; McKenna and Pole 2007, $n = 141$; Pole 2010, $n = 80$; Pole 2005, $n = 20$; Su, Wang, and Mark 2005, $n = 121$). The specialized nature of our studied population – exclusively Canadian citizens who blog independently about politics and publish content at least once a week – and the relatively small size of the active Canadian political blogosphere explain and to a certain extent warrant the limited size of the research sample.

2   These figures do not include 13 bloggers who did not respond to one or more of the questions on federal or provincial identification and federal or provincial membership. Therefore, the subsample of respondents used for this study is 135.

3   There is the logical possibility of members who did not indicate that they identified with a political party but were still members of a political party. Indeed, we had

three respondents who fell into this category. We consider them to be non-partisans. These members may be transient party members, joining to support a nomination or leadership contestant (see Cross 2004); our survey did not probe the motivations behind party membership.

4   Nearly 60 percent of the identifiers posted once or twice a week or less on average, compared with about 51 percent of members and 46 percent of non-partisans.

5   We distinguish between "traditional" and "non-traditional" forms of participation both offline and online. This distinction corresponds with those observed in the literature between conventional forms of political engagement (such as voting in elections, donating, and volunteering in political parties) and alternative forms of engagement (such as activism in interest groups and taking part in public demonstrations or boycotts) (Carty, Cross, and Young 2000; Putnam 2000). This distinction extends to online forms of participation, where citizens may engage in traditional forms of online engagement (by, for instance, participating in partisan activities online) or non-traditional forms of online engagement (such as distributing a viral video for a humanitarian cause or signing an online petition) (Bennett 1998).

13

"We Like This": The Impact of News Websites' Consensus Information on Political Attitudes

J. Scott Matthews and Denver McNeney

Reading a newspaper is a rather lonely way to get your news fix. Exposure to the daily headlines may stimulate political and social discussion, but the act of newspaper consumption itself is a deeply individualized activity, one that isolates readers from news producers and from each other. Although the consumption of television and radio news can, in principle, be a collective experience, the typical assembly of TV or radio news consumers is small – often consisting of a solitary viewer or listener – and even relatively larger groups are likely to consist of a self-selected circle of like-minded family and friends. None of the traditional media, furthermore, offers substantive opportunities for citizen participation in the production of the news. That said, print, television, and radio news outlets do routinely solicit feedback from their audiences. These interactions are, however, generally separated in time and space from the delivery of "the news" – appearing as letters to the editor in the back pages of the typical broadsheet or during a special talkback segment in the closing minutes of a news broadcast. One long-standing feature of the news landscape, talk radio, offers the audience a real-time opportunity to participate in the production of news content (on talk radio in Canada, see Sampert 2012). Yet this exception proves the rule of limited citizen participation in news production: physical constraints imply that only a tiny fraction of the listening audience can take part in the production of any given talk radio program.

The alienation of the citizen from news production and citizens' alienation from each other in the consumption of news are significant from a democratic perspective. First, inasmuch as information environments shape citizen preferences over political alternatives (Bennett and Iyengar 2008), the citizen's role in constituting those environments is a critical concern for representative democracy (Bartels 2003). Second, if, as imagined by deliberative theorists (Gutmann and Thompson 1996), opportunities for citizens to exchange and transform each other's views are a key test of democracy, then the social environment of news consumption is also important. That is to say, it matters greatly whether the citizen consumes the news in isolation or in the company of diverse others who bring differing interpretations to bear on the political events of the day.

Against this backdrop, the appearance of new, Internet-based opportunities for citizens to participate collectively in the consumption *and* production of news content raises an intriguing possibility: the new, online news regime may help engender a more participatory system of political communications. This extends beyond the ability of citizens to author blogs, as described in the preceding chapter. In particular, the widespread introduction on news websites of audience feedback mechanisms – simple polling devices attached to news stories that allow readers to evaluate news content and expose themselves to the aggregated evaluations of fellow readers – may help remedy the democratic deficiencies of the existing regime of political communications. Unlike a letter to the editor in a newspaper or talkback contribution on a TV or radio broadcast, audience feedback mechanisms are temporally and spatially coincident with news content itself: reader evaluation can begin as soon as stories are posted; statistical summaries of audience feedback are updated instantaneously; and, perhaps most importantly, audience feedback on a story is reported alongside the story itself (below or beside regular news content). In contrast to citizen participation in talk radio, furthermore, audience feedback mechanisms are not subject to meaningful physical constraints (there is no limit on the number of people who can evaluate a given story) or editorial gatekeeping, and the costs of participation are little more than the effort involved in forming an evaluation and indicating one's judgment with a mouse click. Seen in this light, audience feedback mechanisms promise to help democratize the production and consumption of news by creating opportunities to shape news content in a manner that is timely, imposes minimal costs on participants, and can involve a virtually unlimited number of citizens (Allan 2006).

The democratic promise of news websites' audience feedback mechanisms depends, however, on the answer to a key question: Do audience feedback mechanisms matter to online news consumers? If the views of one's fellow readers are ignored or simply go unnoticed, then audience feedback mechanisms can neither shape news content nor serve as conduits for collective deliberation.

In this connection, it also bears noting that the influence of "mass collectives" (Mutz 1998) on individual attitudes and behaviours has more often been critiqued for its capacity to frustrate than praised for its potential to facilitate democratic goals. Indeed, the possibility that knowledge of collective opinion threatens the autonomous formation of citizen preferences has worried theorists of democracy at least since Tocqueville (Noelle-Neumann 1974). To some degree, such concerns simply reflect the contrasting premises of representative and deliberative democratic theorists. An important end of deliberation, after all, is precisely to achieve a less isolated, more social process of opinion formation. Still, the image of a mass public dangerously vulnerable to social pressures to conform continues to haunt the study of what psychologists term "majority influence" (Mutz 1998).

Whatever its normative implications, the influence of audience feedback mechanisms on mass attitudes is largely unknown. A modest body of non-Canadian findings on related phenomena suggests that audience feedback can in fact matter. A series of studies of South Korean Internet news consumers, for instance, indicate that the direction (i.e., favourable or unfavourable) of reader comments on a news story can influence associated attitudes (Yang 2008; Lee and Jang 2010). While reader comments are richer in substantive content than audience feedback mechanisms – and, thus, potentially more persuasive – this research nonetheless suggests that Internet news readers are attentive to the views of other readers and may, under the right conditions, be influenced by those views. More directly relevant to the question of the impact of audience feedback mechanisms is research on so-called collaborative filtering technologies, which assign ratings to online news articles based on the number of times an article has been accessed. These studies, which involve US samples, suggest that articles with the highest user ratings are not only more likely to be read, but are likely to be read more closely than other articles (Sundar and Nass 2001; Knobloch-Westerwick et al. 2005). We have good reason to suspect, therefore, that audience feedback mechanisms may be influential for online news consumers, both in Canada and elsewhere. It remains, however, to establish this suspicion empirically.

Before examining the impact of audience feedback mechanisms, we must consider the multiple pathways along which these mechanisms may exert their influence. In this regard, research in social psychology provides useful guidance (e.g., Baker and Petty 1994; Areni, Ferrell, and Wilcox 2000). To summarize the principal conclusions of this research (cf. Petty and

Wegener 1998), it is broadly accepted that "consensus information," such as statistical summaries of the views of others, generally exerts influence on attitudes in one of two basic ways. First, consensus information can directly influence the attitudes of the recipient of that information. Learning that a large majority favours a particular candidate in an election, for example, can induce voters to shift their preferences in the direction of that candidate (Ansolabehere and Iyengar 1994). This is the bandwagon effect well known to political scientists. Second, the presence of consensus information can influence the degree of cognitive effort individuals expend in considering a message or persuasive appeal to which they are exposed. For instance, finding out that a consensus of one's peers takes a certain view on a question of public policy can motivate increased attention to that question and more careful consideration of relevant arguments (Petty and Wegener 1998).

This latter point bears elaboration. A key insight in research on majority influence is that the presence of social consensus may not, on its own, be sufficient to motivate careful consideration of messages that have been endorsed or rejected by majorities. What matters, instead, is the relationship between the social consensus and the pre-existing views of individuals exposed to that consensus. In particular, Baker and Petty (1994) emphasize the role of "source-position balance" in shaping processes of majority influence. They argue that individuals prefer – and, in fact, generally expect – to find the "positions" they hold supported by "majority sources." Conversely, it is unexpected, and possibly threatening, to discover that the majority holds a position one finds inconsistent with one's views or counter-attitudinal. In response, those experiencing such source-position imbalance are motivated to scrutinize arguments more closely in order to account for the discrepancy between their own and the majority's views.

The upshot of this added expenditure of cognitive effort is that the attitudes of those who experience source-position imbalance tend to reflect more faithfully the quality and direction of the arguments to which they are exposed. That is, being motivated to consider persuasive appeals more deeply, those who experience source-position imbalance are more likely to respond to the substantive content of a message and, therefore, their attitudes are more likely to reflect the strength of the message's argumentation.

The above logic yields subtle predictions for the impact of consensus information generally and of audience feedback mechanisms on news websites in particular. We clarify the latter below. For now, it suffices to observe that the influence of audience feedback may depend in part on the political attitudes of the online news consumer and the quality or strength of the arguments tacitly endorsed by the audience.

At the same time, as suggested by the broader findings on majority influence noted above, some readers of news websites may respond directly to the social consensus recorded through audience feedback mechanisms. Put differently, the feedback may serve as a heuristic cue or simple rule that readers can use to inform attitudes relevant to the subject matter of a news story. As in other contexts (see, for example, Axsom, Yates, and Chaiken 1987; Mutz 1998), the reader may take the views of the majority as an indicator of the "correct" opinion. In this way, the audience feedback mechanism becomes a cognitive shortcut to judgment – a low-effort route to opinion formation (Gidengil et al. 2004).

Case Study

METHOD

We investigated the influence of audience feedback mechanisms on Internet news websites with an experiment designed to mimic the experience of reading and responding to the news online. The experiment was delivered to a sample of Canadian university undergraduates. It manipulated the presentation of consensus information in relation to an editorial penned in support of a fictitious proposal for a new airport security measure at Canadian airports. Following the manipulation, we measured attitudes on the airport security issue, with the aim of detecting any effects of the consensus information.

Before discussing the design in detail, we first justify our choice of issue and consider the consequences of utilizing an undergraduate sample. Importantly, we conclude that these features of our study render it a "most likely case" (Eckstein 1971) for the detection of effects of audience feedback mechanisms.

We chose the topic of airport security – specifically a (fictitious) proposal to introduce "FAST," or "Future Attribute Screening Technology," as part of the passenger-screening process at Canadian airports – in order to treat political attitudes on what is known as a "hard issue" (Carmines and Stimson, 1980). Past research has demonstrated that the influence of cognitive heuristics on political attitudes is greater in relation to hard issues – issues that are unfamiliar and technical in nature – than in domains that are salient for citizens on an ongoing basis (e.g., taxation and spending; see

Bailey, Sigelman, and Wilcox 2003; Nicholson 2011). Selecting a hard issue, therefore, should create optimal conditions for the influence of audience feedback mechanisms on attitudes, particularly among those likely to spend less time and energy considering the issue.

In our view, the possible introduction of FAST fits the definition of a hard issue. Subjects could not have pre-existing attitudes toward this specific proposal since, so far as the authors are presently aware, this technology is not being considered for introduction at Canadian airports (although the technology is being examined in the United States). Furthermore, the question is relatively technical and means-oriented and does not manifestly engage the kinds of symbolic considerations, such as political values, that are typically associated with "easy issues" (Carmines and Stimson 1980, 80). Finally, the hard-easy issues distinction aside, the issue raises several conflicting considerations that subjects may find difficult to easily reconcile. This sort of decision-making complexity should, in general, promote reliance on heuristics, including consensus information (Lau and Redlawsk 2006).

For similar reasons, studying audience feedback mechanisms with an undergraduate sample is an especially attractive approach (note that, in Chapter 3, Pénélope Daignault used a similar sampling technique to assess the impact of televised political advertisements). Our sample was 61 percent female and 60 percent ethnically European; had a mean age of twenty-one years; and had a modal pre-tax family income greater than or equal to $110,000 per year (54 percent of participants reported family incomes within this category). Clearly, our sample is not representative of the broader Canadian population. Yet it is this population of young, educated, and affluent citizens that exhibits the highest levels of Internet use, including use of social media and consumption of online news (Pew Research Center for the People and the Press 2008). The effects of audience feedback mechanisms in this group, therefore, are socially and politically consequential. More importantly, inasmuch as they are in the vanguard of new media use (Tapscott 2009), undergraduate students might be expected to be relatively more responsive to audience cues presented on news websites. Furthermore, students are likely to have, in general, lower levels of political knowledge and weaker political attitudes, which should make them more susceptible to the influence of heuristic cues (Lau and Redlawsk 2006).

Our decision, therefore, to focus on a hard issue and to study a high-Internet-use population should maximize our chances of detecting the effects – if any – of audience feedback mechanisms. At the same time, a failure to detect effects in this "most likely" context would suggest, importantly, that audience feedback mechanisms are unlikely to influence attitudes in more commonplace contexts (e.g., on easy issues and in the general population; for a discussion of inference involving most likely cases, see Eckstein 1971).

In March and April 2012, a sample of 359 undergraduates enrolled in second-year political science courses (though not necessarily political science majors) at Queen's University in Kingston, Ontario, was recruited to participate in an experimental study for extra credit. Subjects were recruited via email and invited to participate in a web-based survey, designed to be completed in less than fifteen minutes and concerning attitudes toward airline travel and its presentation in news media. After responding to a series of standard political, social-psychological, and demographic questions, subjects were presented with two short news stories on the topic of air travel, which were constant across subjects, and with an editorial on the FAST system, which contained our experimental manipulations. Subjects were advised that the material had been taken from the website of a major news corporation. In fact, the stories had been generated by the authors. The layout of the articles was carefully designed to mimic the presentation of news on the popular CBC.ca website. Subjects were encouraged to spend as much or as little time as they liked reviewing the articles. Following the presentation of the articles, subjects were asked attitudinal questions designed to capture the study's main dependent variable and additional political and social-psychological items. Subjects were then debriefed and directed to academic publications relevant to the study.

The editorial on FAST explained that the FAST technology "employs a range of non-intrusive sensors to screen travelers for certain behavioural attributes associated with committing violent acts or other crimes." Aside from relaying basic details of the FAST system, the editorial indicated the technology was under testing by the Canadian Air Transport Security Authority. The editorial noted a plausible, if minor, objection to the introduction of the technology (the risk of false positives), but ultimately came down strongly in favour of the proposal, citing its capacity to catch potential threats, reduce "intrusive pat downs and secondary screenings," and render "long security lines during holiday travel seasons ... a thing of the past." As our theoretical framework suggests, the paucity of strong, anti-FAST arguments, and the relative strength of the pro-FAST arguments, is analytically important. We return to this point below.

Subjects were randomly assigned to one of three versions of the editorial on FAST, which differed according to the presence and direction of consensus information. In the control condition, no consensus information was presented; subjects were simply exposed to the editorial. In the two other conditions, however, the editorial was presented alongside consensus information, presented in the form of an indicator of "readers feedback" ostensibly based on positive and negative evaluations from readers of the editorial. Following CBC.ca's conventions for presenting feedback on reader comments, evaluations were expressed as counts of thumbs up and thumbs down and the difference between the two (the editorial's rating). In the positive consensus condition, the editorial received 252 positive (thumbs up) evaluations and 19 negative (thumbs down) evaluations, for a net rating of 233. In the negative consensus condition, the figures were reversed: 252 negative and 19 positive evaluations, for a rating of −233.

To track the dependent variable of interest, we tapped attitudes toward FAST with three measures: "Do you support or oppose the introduction of the new airport screening technology described in the editorial?"; "I would be willing to subject myself to the new airport screening technology described in the editorial." (Agree/Disagree); and "Introducing the new airport screening technology described in the editorial is a good idea." (Agree/Disagree). These measures form a highly reliable index (Cronbach's alpha = .88), which we scale to the (0,1) interval. We describe the construction of additional variables in the course of presenting the results.

RESULTS

To begin, we verified that the random assignment of subjects to treatment conditions was successful by regressing on the experimental factor measures of twenty-three individual characteristics observed in the study (e.g., party identification, need for cognition, age, gender). In none of these regressions do the indicators of the experimental conditions reach the 95 percent threshold for statistical significance.

Second, we investigated subjects' level of attention to the audience feedback by asking, "As you read the editorial on the new airport security screening technology, did you happen to notice whether the editorial was, on average, rated positively or negatively by readers, or did you not notice this information?"[1] Although just 12.6 percent of subjects presented with consensus information indicated that they noticed the readers' feedback, the percentage indicating they noticed positive and negative ratings was clearly differentiated by experimental condition. Positive ratings were reported by 12.5 percent in the positive consensus condition, but by only 2.9 and 5.1 percent in the negative consensus and control conditions respectively. Likewise, negative ratings were reported by 12.9 percent in the negative consensus condition, but by just 4.2 and 1.5 percent in the positive consensus and control conditions respectively. The significance of these results is twofold. On a substantive level, the low level of attention to the consensus information, in spite of its relative prominence, clearly narrows the scope of its potential impact. In this sense, audience feedback did not seem to matter to most of our subjects. Methodologically, however, the sharp differences in perception of the direction of the audience feedback across the positive and negative consensus conditions allow us to conclude that the experimental manipulation was effective, at least for the subset of subjects who were attentive to the consensus information.

So, what was the effect of audience feedback mechanisms on attitudes? Comparing attitudes regarding FAST, as indicated by the support index, across the three levels of the consensus information factor, we found one notable, if counterintuitive, difference: support for FAST was somewhat *higher* in the negative consensus condition ($M = .59$) than in either the positive consensus ($M = .56$) or control ($M = .53$) conditions (though only the difference between the negative consensus and control conditions was significant: $p = .07$). In other words, those subjects exposed to audience feedback indicating a near unanimous *rejection* of the editorial's endorsement of the proposed airport screening technology were actually *more* supportive of that technology than those receiving no audience feedback whatsoever. Furthermore, those in the negative consensus condition were no less (and possibly more) supportive of the technology than those exposed to a near unanimous endorsement of FAST.

These results are superficially puzzling, but recall the diverse effects suggested by our theoretical framework. The simplest possibility sees audience feedback as a judgmental heuristic; for those expending minimal effort in forming evaluations of the FAST system, the balance of reader opinion may be taken to indicate the "correct" attitude. The source-position-balance theory, however, generates quite different expectations. According to this perspective, the influence of audience feedback may depend on a complex interaction between the direction of the social consensus, subjects' pre-existing attitudes, and the strength of the arguments contained in the FAST editorial. Furthermore, these two patterns of influence (i.e., heuristic-based and source-position-balance-based) may operate simultaneously, each applying to a different group of subjects. The heuristic pattern of influence seems most likely to obtain for those subjects who are, in general, unlikely to expend significant cognitive effort in forming social and political attitudes. For other subjects, the influence of audience feedback seems likely to depend on the compatibility of consensus information with pre-existing attitudes, per source-position-balance theory.

To explore these possibilities, we divided our sample according to indicators of cognitive effort and pre-existing attitudes relevant to the airport screening proposal. We relied on a measure of "need for cognition" (or NFC; Cacioppo and Petty 1982) to differentiate those subjects likely to expend a high level of cognitive effort in assessing FAST from those who are less likely to do so. Need for cognition is a widely employed construct in social psychology and, increasingly, in political science (e.g., Cindy Kam 2005) that taps individuals' general inclination to take part in and enjoy cognitive activity (Cacioppo and Petty 1982). Critically, as compared with those low in NFC, individuals high in NFC have been shown to be more likely to engage in time- and energy-consuming thought concerning issue-relevant argumentation, rather than to rely on simple heuristics or cues. Our measure of the concept consists of two components of the standard NFC battery, adapted for the web-based survey mode.[2] To identify subjects likely to possess pre-existing attitudes favourable or unfavourable to FAST, we relied on an indicator of party identification (PID). Although party identification is, at best, an indirect measure of the relevant pre-existing attitudes, the policy orientations of Canadian partisans lead us to anticipate substantial differences across partisan groups in evaluations of FAST. In particular, we would expect Conservative partisans, who tend to attach the highest priority to and take the most ideologically conservative positions on issues of border security, to find the introduction of FAST at Canadian airports relatively more attractive than do other Canadians. This assumption is supported by a regression (not reported) of the index of FAST support on indicators for Conservative, Liberal, New Democratic, and Green Party partisanship (reference category: all others): Conservative partisans were more supportive of FAST than other partisans and non-partisans (difference significant at the 95 percent level).

Crossing these two variables – need for cognition and party identification – yields the fourfold division of our sample represented in Table 13.1. In

Table 13.1

Predicted changes in FAST support

	Other partisans, Non-partisans	Conservative partisans
Low need for cognition	Positive: Increase Negative: Decrease	Positive: Increase Negative: Decrease
High need for cognition	Positive: Increase Negative: No change	Positive: No change Negative: Increase

NOTE: Main cell entries: Direction of consensus information: change in FAST support, relative to the control condition.

the rows, "high need for cognition" subjects – those at or above the sample median – are distinguished from "low need for cognition" individuals. In the columns, Conservative partisans are separated from all others. Each cell shows a predicted change in support for FAST, based on the conclusions of the majority influence literature, and relative to support in the control condition, given exposure to positive and negative consensus information.[3]

For low-NFC subjects, we expected attitudes toward FAST to largely mirror the direction of the audience consensus. In other words, low-NFC respondents' attitudes ought to be more favourable in the positive consensus condition than in either the control or negative consensus conditions (i.e., support should increase in the positive consensus condition) and vice versa for the negative consensus condition. Such a pattern would be consistent with online news readers' use of audience feedback as a judgmental heuristic. The predictions are the same irrespective of party identification; we would expect Conservative and non-Conservative partisans to respond similarly to consensus information if it is being utilized as a simple shortcut to attitude formation.

For subjects high in need for cognition, however, we anticipated a more complex pattern of effects. Being less likely to rely on simple heuristics, these subjects' responses to consensus information should depend on pre-existing attitudes relevant to FAST and on the quality of the arguments the editorial contains. High-need-for-cognition Conservative partisans, whose pre-existing attitudes should be favourable to FAST, should not be much affected by positive consensus information: the audience feedback affirms a conclusion these subjects are likely to have already reached. The balance between source and position, in other words, implies that these subjects will have little additional motivation to reflect on the arguments presented in the editorial (per Baker and Petty 1994). For the same reason, high-NFC "other" partisans and non-partisans, whose pre-existing attitudes should – relative to Conservative partisans – be unfavourable to the airport screening technology, should not be affected by negative consensus information: the audience's rejection of the FAST system is in balance with these subjects' rejection of the technology and, therefore, the consensus information does not supply new motivation to carefully consider the editorial's arguments.

The remaining two groups of high-NFC subjects, however, should find added motivation for effortful cognition in their exposure to consensus information. High-need-for-cognition Conservative partisans in the negative consensus condition were expected to resist the social consensus: source-position imbalance should motivate closer scrutiny of the editorial's arguments, and the availability of strong, pro-FAST arguments in the editorial should reinforce their existing attitudes. The result should be a boomerang effect: that is, high-NFC Conservatives in the negative consensus condition should be *more* favourable toward FAST than are their counterparts in the control or positive consensus conditions. High-NFC non-Conservatives in the positive consensus condition should also find themselves in a state of imbalance and, thus, motivated to carefully reflect on the editorial's endorsement of the airport security proposal. The resulting elaboration should push these subjects, however reluctantly, in the direction of support for the FAST system, given the strength of the editorial's arguments in favour of the proposal.

The experimental results provide modest support for these expectations. We took account of the moderating effects of need for cognition and partisanship by estimating an OLS regression of the FAST support index on the consensus information factor, along with two- and three-way interactions between this factor, need for cognition, and party identification. (To enhance the precision of the estimates, we also include controls for income and gender).[4] Coefficient estimates for this model are reported in Table 13.2. Figure 13.1 graphs the predicted marginal effect on the support index of exposure to positive and negative consensus information (relative to the control condition) for each combination of need for cognition and partisanship represented in Table 13.1. The capped bars indicate 90 percent confidence intervals; significant effects are those with confidence intervals that exclude zero.

Table 13.2

Regression results, FAST support index

	β	(se)
Positive consensus condition	0.0322	(0.0511)
Negative consensus condition	0.0339	(0.0496)
High need for cognition	−0.0232	(0.0456)
Conservative party identification	0.0473	(0.0690)
Positive consensus × High NFC	0.0155	(0.0663)
Negative consensus × High NFC	0.0253	(0.0674)
Positive consensus × Conservative PID	0.0154	(0.1001)
Negative consensus × Conservative PID	0.0344	(0.1048)
High NFC × Conservative PID	0.0794	(0.0877)
Positive consensus × High NFC × Conservative PID	−0.0517	(0.1302)
Negative consensus × High NFC × Conservative PID	0.0485	(0.1419)
Gender†	−0.0622	(0.0247)
Income	0.1276	(0.0399)
Constant	0.4582	(0.0505)
Observations	359	
R-squared		0.1099

NOTE: OLS estimates.

† Woman = 1.

As Figure 13.1 indicates, just one of the predicted marginal effects reached conventional levels of statistical significance: among Conservative partisans who are high in need for cognition, exposure to negative consensus information increases support for FAST by .14 units (significant at the 90 percent level). The effect is substantively significant: relative to the level of support in the control condition, exposure to negative consensus information increases support among high-NFC Conservatives by more than one-quarter. The pattern fits the anticipated boomerang effect; that is, high-NFC Conservatives appear to have reacted to the (presumably) counter-attitudinal consensus information by more closely scrutinizing the editorials' arguments. Put differently, the counter-attitudinal negative consensus information would seem to have motivated high-NFC Conservatives to process more assiduously the editorial's strong advocacy for FAST, which in turn reinforced these subjects' favourable evaluations of the airport screening system.

FIGURE 13.1

Effects of consensus information by need for cognition and partisanship

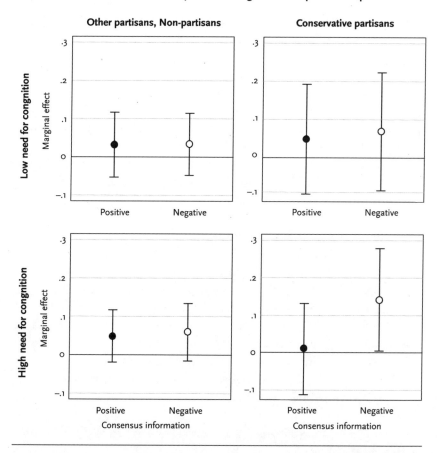

None of the remaining marginal effects graphed in Figure 13.1 is statistically significant. For high-NFC Conservatives exposed to positive consensus information, and for high-NFC non-Conservatives exposed to negative consensus information, this comports with broader findings on majority influence: the consensus information should appear attitudinally congruent – source and position should be in balance – and therefore have little effect. Less explicable is the non-effect of positive consensus information among high-NFC non-Conservatives. One possibility is that these subjects were not sufficiently opposed to the airport screening proposal to be motivated by a positive consensus in the system's favour. In other words, high-need-for-cognition non-Conservatives exposed to the positive consensus may not have experienced source-position imbalance to the same degree as did high-NFC Conservatives exposed to the negative consensus. The most surprising results are those for low-NFC subjects; we observed no effect of either type of consensus information in this group. The views of these individuals, we surmised, should move in tandem with the direction of the consensus information to which they were exposed. Yet their evaluations are undifferentiated by exposure to vastly different kinds of audience feedback. It is imaginable that the strong, pro-FAST arguments in the editorial simply overwhelmed these respondents, but this is theoretically unlikely: low-NFC respondents are unlikely to process argumentative content deeply, and if they did, we would expect the extent of processing to be differentiated by the degree of source-position imbalance.

Political Communication in Canada

The results in this study suggest a subtle view of the political significance of the new, more participatory regime of online news. Indeed, its potential to render the political communication system more participatory is among the most novel – and from a democratic perspective, most attractive – features of the new world of online news. Putting aside concerns about threats to citizen autonomy in preference formation, such technology as audience feedback mechanisms on news websites holds the potential, in principle, to involve citizens both in the production of news and in lively collective conversation – even deliberation – about the meaning of political developments. Is that potential being realized?

Our investigation allows both a "glass half empty" and a "glass half full" answer to the above question. The participatory glass is half empty if we focus on the absence of effects of consensus information among those low in need for cognition. This group, it has been shown (Petty and Wegener 1998), typically processes messages heuristically: exerting only a minimum of cognitive effort, they are influenced by features of messages that are peripheral or extrinsic to messages' substantive content. The presence of a clear social consensus should suggest an efficient route to attitude formation for such individuals. Yet, in our study, such information had no effect on low-NFC subjects. Why? It is instructive that just 12.6 percent of subjects reported noticing the audience feedback information presented with the editorial, a proportion that is not differentiated by need for cognition (analysis not reported). Put differently, almost nine out of ten subjects failed even to detect the consensus information with which they were presented, even though that information was quite prominent: clearly labelled, expressed in two different ways (i.e., thumbs up/down and a summary rating), and placed next to the first paragraph of the editorial. The finding suggests that awareness of audience feedback on real-world news websites, where such information is typically less prominent, may be even lower. It is also worth bearing in mind, as discussed above, that the present study should have created optimal conditions for the use of heuristic information, including consensus cues, given its focus on the hard issue of airport security measures and the characteristics of our undergraduate sample. Attention to news websites' consensus information in the general population and in regards to the "easier" issues more central to political conflict may, therefore, be even lower. In this light, the participatory news regime seems more monologue than dialogue.

On the other hand, the participatory glass looks half full if we focus on effects among Conservative partisans high in need for cognition. Such individuals are expected to attend to the intrinsic, argumentative content of political communications – to engage in systematic evaluation of the substantive merits of the messages to which they are exposed. In our study, it seems that high-NFC Conservatives conformed to expectations. Indeed, the effect of (negative) consensus information in this group was large. Learning that a large majority of readers held views contrary to their own, we reason, motivated high-need-for-cognition Conservatives to think more carefully about arguments in favour of the proposed airport screening technology. The arguments these subjects found, of course, were disproportionately strong, and so this careful thinking merely *reinforced* subjects' initial opinions. In the presence of a more balanced set of arguments, however, source-position imbalance may induce attitude *conversion*. This suggests that, under the right conditions, the dialogical potential of the online news regime may be realized.

Ultimately, our analysis suggests, the democratic potential of audience feedback mechanisms – their capacity to produce consequential citizen participation in news production and meaningful collective deliberation over news content – depends on the particular mix of political and cognitive characteristics present in the population. Upward of 40 percent of Canadians typically confess to enduring partisan commitments of varying strength (Gidengil et al. 2012). The distribution of need for cognition, furthermore, is fairly uniform, with almost half of Canadians reporting levels above the national median (Canadian Election Study 2011). This implies, generalizing from our results, that just one in five Canadians are in a position to respond to audience feedback of the right form – that is, consensus information that implies that a given citizen is on the "wrong side" of public opinion. Conversely, a large majority of Canadians seem unlikely even to notice consensus information. And many of those who notice it are unlikely to be moved, especially if – as seems logically necessary – audience feedback mechanisms tend to report social consensuses that are congenial for most of the population.

Of course, audience feedback mechanisms on news websites are just one part of the broader online news regime. Whether our findings generalize to other domains is a question for future research.

NOTES

1   The question also included this clarification in parentheses: "The editorial's rating was indicated by counts of the number of 'thumbs up' and 'thumbs down' the editorial received."

2   Subjects were asked to indicate whether the following two statements were characteristic of themselves: "I would prefer complex to simple problems"; "Thinking is not my idea of fun." We chose the components based on a factor analytic study of the standard battery reported in Cacioppo et al. (1996). Our measure includes the highest loading component and, to offset acquiescence bias, the highest loading of the remaining components that were phrased in the opposite evaluative direction. In the analysis, we compute an additive index of the two components.

3   These expectations also depend on an implicit, but highly plausible, assumption: that subjects in the control either did not consider the state of the social consensus or imagined that opinion on the issue was roughly divided.

4   These controls were the only two that were significant in a larger regression (not reported) that also included age, religiosity, air travel, and political trust.

14

Political Communication and Marketing in Canada: Challenges for Democracy

Alex Marland, Thierry Giasson, and Tamara A. Small

This book has examined the mediation of political discourse in Canada from a positivist approach of describing "what is" rather than as a normative "what should be" critique. Empirical data have been presented in a manner that encourages readers to consider evidence and arrive at their own informed conclusions. It is our view that scholarly judgments must be balanced with an appreciation for the realities of Canadian political institutions, political news media, and citizens. The practice of political communication is evolving in response to technological change, and the evidence presented in this book demonstrates that it diverges from academic ideals.

The last transformative era in political communication occurred with the emergence of television in the mid-twentieth century. That visual medium radically changed how politicians communicate, how political events are reported, and how audiences consume political news. When Web 1.0 surfaced in the 1990s, citizens had opportunities to become content creators, but the one-way flow of political information persisted in websites. It was the diffusion of Internet technologies and advent of Web 2.0's two-way communication via online forums, blogs, social networking, microblogging, and smartphone applications that has involved more dramatic change. Canadian political actors are continuing to respond to this new and changing media environment.

It is our assessment that in many respects the response by political institutions in Canada could be described with the epigram "Plus ça change, plus c'est pareil." This is because existing communication practices tend to be adapted to new platforms and new environments. Negative political advertising, image management, and jockeying for message control with the press are among the many tactics discussed here that have always been present in Canadian politics. Nevertheless, we feel that the increasingly fast, unmediated interactions among political elites and citizens mark a transformative period. Communication by political parties is becoming more strategic, more succinct, more intense, and more pinpointed. Professional journalists no longer control all the means of mass communication, and are competing with partisans and citizens to circulate political information. There is more attention to dramatic events, to personalities, and to certain policy issues over others. Canadians have growing opportunities to connect with interest groups. They can share their political views with strangers around the world and instantly render public judgment on news stories. This is an interesting time to be studying political communication, and although we are unable to arrive at a firm conclusion as to whether this is a net positive or a net negative for Canadian democracy, we lean toward the former. Our view is that changes in communication technologies that are enabling Canadians to interact in an online public sphere will improve the responsiveness of political elites.

Political Communication in Canada has attempted to shed light on three unifying research themes. First, we have sought to understand current trends in Canadian political communication. Second, we have explored what kinds of effects the new political communication environment is having on Canadians. Finally, we have considered the implications for Canadian democratic discourse. In this chapter we put the findings in context before proceeding to further illustrate how changes in elite behaviour and technology are increasing the strategic precision of political communication in Canada.

Communication by Canadian Political Institutions

Every political communication action is geared, directly or not, toward citizens. Newscasts, blogging, electoral ads, party leaders' tweets, government communications, and grassroots advocacy campaigns all share the same finality: to influence how Canadians think and act politically. Chapters in the first part of this book applied political communication theories to Canadian political parties' use of permanent campaigning, television advertising, brand-image management, and social media. Collectively, they indicate that Canada is not immune to the trend in Western democracies for political actors to use strategic marketing and communications tactics. A broad theme that connects much of the book's content is the notion of the permanent campaign. In Chapter 2, Anna Esselment examined how after federal elections the Conservative government has continued to battle with its opponents. This includes a relentless effort to stay on script, the polarization of messaging, the use of market intelligence, divisive partisan advertising, and the centralization of decisions. As a result, communication by the Government of Canada is increasingly looking like campaign communication, as does political parties' inter-election behaviour.

Advertising, information subsidies, and branding are among the ways that parties aim to define and control their public images and that of their opponents. Democratic politics inherently involves a contrast between competing ideas and candidates. Practitioners are cognizant that in a thirty-second television spot, a photograph, or even an election manifesto, there is a limited opportunity to make a desired impression on their audiences. Positive and negative messages must be simple, truthful, and memorable, and must be repeated over and over again. The melding of spokesperson personality and policy positions across many media platforms means that it is essential to unify all aspects of the party message.

Many of these concepts can be found in Pénélope Daignault's measurement of whether Canadian political TV advertising results in positive or negative judgments. In Chapter 3, she found that negative election ads are not as persuasive as their sponsors may intend. Her research indicates that comparative spots trigger more mental processing, and that this form of communication may foremost reinforce existing partisan leanings. These themes are also present in Alex Marland's discussion in Chapter 4 about the PMO's "photo of the day" tactic, which provided clues about how Conservative strategists wanted to position Prime Minister Harper's brand image. More significantly, it is an example of how government communication is entwined with party communication and how political actors are seeking to exploit a changing media landscape. We can also see it in the observations of Jared Wesley and Mike Moyes in Chapter 5. Their assessment of the brand-image management of the NDP noted a process of inoculation, moderation, and simplification that included a strategy to deflect public attention from the party's negative characteristics – its extremes and complexities. Trends toward message simplicity, obsession with image, and fewer promises are evidently trickling not only from the United States and United Kingdom to Ottawa, but also to the provinces.

Tamara Small's assessment in Chapter 6 of how party leaders used Twitter in the 2011 election campaign is further evidence of a preference for political actors to control the narrative. Despite the great expectations associated with the potential for e-democracy to engage citizens in politics, to date, the Internet has not revolutionized Canadian politics. Social media is an important tool, but it, like other Internet communication technologies, is used by mainstream political parties foremost to bypass news filtering with messages that are also promoted on other mediums. Social media does, however, provide an opportunity to level the communications playing field for small organizations and for Canadian citizens who access such platforms.

Canadian News Coverage of Politics

The second part of the book explored changes in political media relations and the content of political news coverage. Chapters examined the relevance of the press gallery, how newspapers cover an election campaign, the media's interest in personality politics, and how policy issues are framed in the news. Unifying these analyses is the tug-of-war battle between the media and parties, which has brought a change of tone in press coverage. The reduced media availability of the prime minister and partisans' preference for controlled events are part of the permanent campaign. In the past, the relationship between government decision makers and the journalists covering them has at times been too comfortable, leading to media reports that were propagandist in nature. This was the quid pro quo of granting reporters access to political elites and insider information. Now the relationship is often too acrimonious, as indicated in Chapter 7 by Daniel J. Paré and Susan Delacourt, making it difficult for the media to gather, analyze, and report on the details of Canadian government and politics.

The accessibility and quality of available political information is important because Canadians are influenced by news. Media coverage of politics helps define the public's political agenda, a notion that most of this book's authors have identified. The priorities of voters during an election campaign are very much conditioned, as Elisabeth Gidengil's chapter on the 2006 federal campaign attests, by the importance that news organizations dedicate to specific issues. The more coverage an issue receives, the likelier the possibility that the issue will become a priority to electors. Furthermore, intensity of coverage and the way that some issues are depicted – or framed – in the news can impact how Canadians relate to policy proposals to address societal issues. The case of welfare policy framing presented in Chapter 10 by Adam Mahon, Andrea Lawlor, and Stuart Soroka indicates that the media cast some social issues in rather negative interpretative schemas. According to framing theory, the news media may be efficient in relaying these interpretations to the population, which in turn may evaluate the problem and those associated with it (political actors and other citizens alike) in an unfavourable way. Frames are sometimes media-generated, but they are often government- or party-generated. As Entman's (2004) cascading activation model revealed, political actors can impose a frame to the media which – if certain conditions are present, such as the absence of opposing frames and the cultural convergence of the imposed elite frame – in turn relay it to the population. Again, this is a key question, as these frames help citizens understand how social issues should be addressed with public policy initiatives. However, frames withhold certain information from audiences, and news is presented with a bias that can influence mass opinion. Frames serve ideological purposes, and they can play a strong role in shaping public support on policy making.

Political news is pervasive and electoral news is, for the duration of a campaign, omnipresent. There has never been so much and so much diverse political news available to Canadians. This is not to say that Canadians are more engaged in the development of public policy. The plethora of media choices, the dominance of strategic framing in news coverage of politics (see Giasson 2012), and the attention to politicians' private lives and personalities (see Chapter 9 by Mireille Lalancette) mean that citizens have access to more political information but are not necessarily better informed. Many chapters in this book indicate the potential influence of the Canadian news media on setting public priorities. They point to the media's ability to orient citizens' evaluation of leaders' personalities and to impose interpretative frames. But it remains unclear just how effective the news media is in helping Canadians make enlightened political and electoral choices. Further investigation on normative issues such as the quality and usefulness of political news coverage for citizens should be at the top of Canadian political communication scholars' research agendas.

Political Communication and Canadian Citizens

The final part of the book has explored the extent to which the new communication environment is resulting in a more informed, engaged, and/or cynical citizenry in Canada. Contributors looked at how non-profits communicate, the nature of political blogging, and how audience comments posted on news websites can shape political attitudes. These chapters are unified in a finding that Internet communication technologies are providing citizens with more ways to engage in Canadian political discourse. It is here where we find limited positive evidence of the transformative aspect of the new environment on Canadian democracy.

Georgina Grosenick's assessment of the communication by Canadian non-profit organizations shows that the Internet has provided an important tool for more effective advocacy, mobilization, and fundraising. However, its availability is not a magic wand, and she observes that non-profits "have yet to effectively employ new and social media." Such groups are often shackled by funding problems, and they compete for media attention not only with each other – there are approximately 161,000 non-profit organizations in Canada, she notes – but with the competitive nature of party politics, the agenda-setting dynamics of the news media, and the plights of individuals. Until non-profits change their "shotgun approach" to political communication, many will not successfully avail of the potential of new technology.

Blogging is another way that citizens are using new technology for political purposes. One indication of malaise toward political news coverage is the increasing number of Canadians who flock toward citizen-generated communication. Political blogs, independent media, and citizen journalism initiatives are gaining followers and readers. Citizens are now active receptors of political communication, both consumers and producers of political messages. This creates a dynamic space for exchange and debate. However, this space is not accessible to all Canadians, and there are some gaps in information gains. As Thierry Giasson, Harold Jansen, and Royce Koop's chapter on political bloggers shows, on one side there is a minority of politically active and interested online users who are the information-rich, and on the other is the vast majority of Canadians who do not actively seek political information online. Citizen-driven political information, whether on the Web or offline, still benefits only a small community.

Social media has transformed the way citizens access political information and interact with political communication. Online news is now a participatory experience. Blogs, social networking sites, file sharing sites, and microblogging tools have given more power and influence to citizens in the way political communication is produced and transmitted. These new mediums bring back the notion of a shared experience of consumption. Feedback mechanisms that are built-in features of these tools, such as widgets indicating numbers of likes and friends a person has on Facebook, allow citizens to share with their followers what they read, appreciate, or deem valuable and credible. In Chapter 13, Scott Matthews and Denver McNeney indicated that audience feedback on news websites might not have a significant impact on how the majority of citizens assess the validity of a news item. Rather, these opportunities to share impressions with strangers hold "the potential to involve citizens both in the production of news and in lively collective conversation – even deliberation – about the meaning of political developments." We tend to agree with them that new technology is providing more forums for democratic discourse.

Ultimately, citizens' opinions are not blank canvasses that react instantaneously to political messages they are exposed to. Rather, Canadians have predispositions and attitudes that limit the persuasive power of political communication. Negativity in advertising or contradictory information carried on news sites stimulates higher levels of cognitive resources and deliberation needed by individuals to block information that counter their predispositions. This constant activity loops back to pressures on political practitioners to engage in a permanent campaign.

The Practice of Political Communication in Canada

These observations must be considered in the context of a strategic and sophisticated political communication marketplace. As mentioned, *Political Marketing in Canada* (Marland, Giasson, and Lees-Marshment 2012) established that there is a preference in Canadian politics for research data to be used to inform communications decisions; that is, political strategists are more interested in figuring out how to push their own ideological agendas than they are in changing in response to public preferences. As a consequence, practitioners increasingly craft communication according to the different publics that they wish to reach – a much more specialized, limited, and circumvented public. Political communication emanating from institutions is said to be micro-segmented and hyper-targeted, using themes, arguments, and even wording that resonate with certain electors and not others. Applying W.R. Smith's (1956) market segmentation analogy to politics, whereas Canadian political parties used to be concerned with securing a layer of the electorate cake, they are increasingly interested in "wedge-shaped pieces" (5). The segmentation approach consists of

> viewing a heterogeneous market (one characterized by divergent demand) as a number of smaller homogeneous markets in response to differing product preferences among important market segments.

It is attributable to the desires of consumers or users for more precise satisfaction of their varying wants ... attention to market segmentation may be regarded as a condition or cost of growth. Their
core markets have already been developed on a generalized basis to
the point where additional advertising and selling expenditures are
yielding diminishing returns. Attention to smaller or fringe market
segments, which may have small potentials individually but are of
crucial importance in the aggregate, may be indicated. (W.R. Smith
1956, 6-7)

In such works as *New Media Campaigns and the Managed Citizen*
(Howard 2006), *Harper's Team* (Flanagan 2009), and Chapter 5 of *Political
Marketing in Canada* (Turcotte 2012), we are reminded that political organizations invest in communication with key strategic goals in mind: to
gather citizens' personal information through data mining and to use this
intelligence to inform hyper-targeted personalized communication campaigns. Though the concept of market segmentation is not new, innovations in computer software have enabled sophisticated data analysis and the
clustering of segments of the electorate. This allows political strategists to
identify the floating voters who are most likely to respond favourably to specialized political appeals. Narrow messages are being communicated via direct marketing and via the media channels that consumption data indicate
are most frequented by the targeted electors. This is a rational way for communicators to maximize limited resources and gain a competitive edge. However, focused appeals to subgroups means that electors who are deemed
to be less important are left out. This is not just people who are profiled to be
opponents but anyone who is considered unlikely to vote, donate, or be
otherwise politically active. As well, as a number of chapters have shown,
there is an emphasis on communicating negative messages in an attempt to
create a partisan wedge.

The narrowcasting of messages is now widely practised by Canadian
political parties who follow the intermingled concepts of relationship marketing, database marketing, and direct marketing (for conceptual nuances
see Johansen 2012). Gone are the days of political actors collecting basic
information that is available in a public phone book for the exclusive purposes of getting out the vote on Election Day. A permanent campaign phil-
osophy means that they constantly update databases of elector information
for voter profiling, identification, and communication. Liberalist, NDP Vote,

and the Conservative Party's Constituency Information Management System (CIMS) store information obtained from the official list of electors that is supplemented with data compiled from telephone directories, from voter interactions with a party or candidate, with poll-by-poll voting results, with opinion surveys, from event attendance, and potentially private sector list providers (but see Flood 2013). The most common way of generating information about electors is when they willingly provide it in response to requests on political party websites. They are not necessarily told that this will cause them to be added to a party listserv that sends out messages drawing attention to a political cause and calling for action, normally in the form of a donation to the party to support its work. As with most listserv messages, there is an option to unsubscribe, but this is not the same thing as the party permanently deleting all records from its database.

Canadian political parties tend to be guarded about releasing details of their proprietary software; doing so would risk nullifying competitive advantages. The Conservative Party requires that an application form be vetted before a password will be given to gain access to a secure CIMS site (Conservative Party of Canada 2013). Less is known about NDP Vote; in early 2013 there was no mention of it on the NDP website and, as one political blogger has observed, the database "has slid under the radar" (Bumstead 2012). Conversely, the Liberalist website housed a plethora of information about their database software, thereby pitting the idealism of democracy against the realities of party warfare. Further research is needed to explore whether these practices are contributing to a more informed, engaged, and connected citizenry or whether this reinforces partisan ideologies, unfairly depicts opponents as wicked, and agitates a cynical electorate.

Available information about these parties' database, direct, and relationship marketing is connected to all of the themes that have been explored in *Political Communication in Canada*. The first of four observations is that the news media are paying more attention to the evolution of voter identification and political strategy (as per Giasson 2012). As CBC News has reported:

> Particularly valuable now [to Canadian political parties] are digital contact information such as email addresses or Twitter accounts, to allow for very inexpensive direct messaging. Today, the limits to the information collected are endless, including credit card information (useful for financial donations), birthdays and anniversaries (useful for customized greetings), religious faiths and ethnicities (useful

for targeting holiday or other custom messages, including ones in native tongues), or even details about someone's education history (public or religious school supporter? post-secondary grad?) or work life (union member? small business owner?) for targeting a specific appeal in the future. Campaigns also may track what voters think about issues and policies, such as which voters support ending the gun registry or legalizing marijuana. (McGregor 2012)

After the media explain to citizens how voter ID databases work, there is inevitably debate about the partisan purpose, the democratic value, and the potential for abuse. Practitioners argue that data analytics allow campaigns to personalize communications with citizens, to reduce assumptions about cohort stereotypes, and to engage with people on an individual level (Flood 2013). Critics point out that data collected by local volunteers become the property of the party executive, which increases the centralization of political communication and the potential for party discipline while reducing the autonomy of local representatives (Canadian Press 2012). Inputted data about electors can be used for fundraising to enable the creation and circulation of inflammatory but truthful information in negative ads that, as Daignault remarked in Chapter 3, beg regulation. Parties can draw on database information to select who receives negative messages. This has been widely linked to the robocalls scandal whereby automated phone messages in the 2011 federal election falsely stated that polling stations had been moved, as well as to post-election phone calls in a Montreal riding falsely stating that the Liberal MP was resigning, both of which allegedly benefited the Conservative Party (Berthiaume 2011). Such practices have led the Liberal Party (2011, 30-31) to lament CIMS's role as a "modern political communication weapon" that "has enabled the advantages of technology to be efficiently leveraged in a relentless drive to (1) control the political agenda and message, (2) pursue perceived political advantage over sound public policy, and (3) quash all internal debate and dissent on any of the foregoing." The democratic value of voter databases, as with so many of the tools discussed in this book, is therefore dependent on how political actors choose to use them.

That the greatest volume of attention to the existence of party databases appears to have been paid when the robocalls scandal topped the news cycle is consistent with Gidengil's conclusion in Chapter 8 that agenda setting is more likely when the media's attention is drawn to a "dramatic event."

Furthermore, such information management has become an organizational centrepiece for political parties, which Esselment warned in Chapter 2 risks fuelling political cynicism and voter alienation. Because all major federal parties maintain databases, it falls to independent bodies such as Elections Canada or the Office of the Privacy Commissioner of Canada to investigate the lack of applicable privacy laws and the potential for breaches (Canadian Press 2007; Bennett and Bayley 2012). However, there is some irony that the media express concern about citizens' privacy but are interested in the private lives of politicians, as Lalancette described in Chapter 9.

Our second observation is that the reduced gatekeeping power of news organizations is illustrated by the changing nature of content production and distribution in the digital age. Party databases are used to inform the circulation of partisan information that bypasses the filter of the mainstream media, which Marland concluded in Chapter 4 is contributing to the convergence of party, personal, and government communication. However, there is a tit-for-tat. Though the Conservative Party may restrict access to CIMS, a veteran political journalist has used the Web to share an internal party CIMS training file (Delacourt 2012); an online newspaper has revealed details of a Conservative campaign school's CIMS training (Pullman 2012); and a political blogger has analyzed party spending declarations to speculate that in 2010 funds were likely used to update CIMS software (Pundits' Guide 2011). This is evidence of how the reduction or elimination of editorial filters allows a wider flow of information. Arriving at a judgment of whether this is good or bad is complicated by the possibility that such content may be highly subjective. It is possible that party databases may draw false information from questionable sources and that content is often posted by authors who have their own political agendas. The blogosphere and social media are a complement to a professional news media, not a substitute.

With the explosion of available political information, it falls to citizens to discern what is real, what is truthful, and what is factual. In Chapter 7, Paré and Delacourt observed that although the mainstream media have been profoundly impacted by new technologies, there is a pressing need for the filtering and mediation that only political journalists can deliver. Conversely, in Chapter 10, Mahon, Lawlor, and Soroka established that the quality of political information is an indicator of the state of democracy, and that biases are present in the Canadian mainstream media's treatment of certain policy issues. The need for balanced information is also found in research by Giasson, Jansen, and Koop in Chapter 12 that identified the pervasiveness of partisanship in a political blogosphere that is characterized by "hyper-citizens." Likewise, in Chapter 13, Matthews and McNeney reported that websites provide opportunities for citizens to share comments, but these are also marked by partisanship.

Third, while much is made of Web 2.0's ability to connect parties and electors, the evolution of voter information programs demonstrates how the Internet improves internal party cohesion. Political elites can easily exchange information with party activists and grassroots workers across the country. Conservative riding offices and local party campaign headquarters access CIMS using an encrypted virtual private network while parliamentarians and party staffers connect via their office computers on a terminal server client (Conservative Party of Canada n.d.). The software allows party workers to input and share information about constituents. They can use it to send mail, and during elections can better manage who has requested a lawn sign. Records about each elector identifies their contact information, ranks the level of their current and lifetime support for the party on a thermometer scale from −15 to +15, and provides details of their political activism. Users are able to generate reports for door-to-door and telephone canvassing and use universal product code (UPC) barcode symbols to track updated information. All of this is used to customize their communication with Canadians.

Whereas in 2007 the CIMS was a tool to store information about electors, the 2013 edition of Liberalist reflects evolutions in communications technology by allowing users to make scripted calls and input caller responses into a virtual phone bank. Users are able to send email "blasts," print canvass sheets and form letters, create survey questions for scripts, record robocall messages, plan public events, and manage volunteers. But political parties seeking to form the government are not in the business of championing democratic idealism unless they believe that it will pay off in votes. The purpose of Liberalist is unabashedly to provide supporters with a mechanism "to prepare for the next election" (Liberal Party of Canada 2013a). Local use of the software is supported through online training to members of its riding associations, with course names such as "data entry made easy," "finding your voters and smart canvass tools," "managing events and donors," and "robocalls and analyzing your data" (Liberal Party of Canada 2013b). This adds evidence to Small's conclusion in Chapter 6 that Canadian political parties prioritize the use of new media for their own purposes rather than for the broader democratic good. Even so, Liberalist is foremost a "voter identification and relationship management system" which, as with CIMS and NDP Vote, is intended to improve the ability for the party to engage with electors throughout Canada (Liberal Party of Canada 2013a). This is where database marketing and direct communication cross into the realm of relationship marketing albeit in a manner that foremost engages a partisan segment of the electorate.

Fourth, these tools further signify that Canadian practitioners draw on the expertise of American political communicators, and that both are connected with their British counterparts. CIMS is similar to the Republican Party's Voter Vault, which has also been a model for UK Conservatives (Seawright 2005). Liberalist was designed by the same Washington-based technology firm that developed the VoteBuilder software used by the Obama campaign and the Democratic National Committee (Liberal Party of Canada 2013d), and that also designed Connect for the UK Liberal Democrats (Liberal Democrats 2013). Since little information is available about NDP Vote it is unclear to what extent US and UK communicators have had direct involvement with that party. Nevertheless, given the influence of the New Labour model described by Wesley and Moyes in Chapter 5, it stands to reason that the NDP has drawn inspiration from Labour's Contact Creator, which has been used to manage direct marketing with electors (Foster 2010, 49). Canadian political parties' commitment to such tools is additional evidence of the pervasiveness of the permanent campaign.

Despite the media's focus on the negative, many positives associated with these databases may be highlighted. They are a relationship management tool for political parties and they increase the ability for parties to communicate directly with interested citizens. The Obama campaign was credited for using such technology to instill a more active culture of participatory democracy (Foster 2010; Baarda and Luppicini 2012) and the transparency associated with Liberalist suggests that the Liberal Party is attempting to imitate the Obama approach. Database marketing is also a model that non-profit organizations stand to emulate and some, such as World Wildlife Fund Canada, are using it to engage supporters (Hart, Greenfield, and Johnston 2005, 201). The marriage of fundraising and relationship building via database marketing can address some of the problems faced by non-profits that Grosenick identified in Chapter 11. As she observed, Canadian non-profits "must find ways to centralize public communication" and to "more effectively facilitate their communication activities." Although Canada's non-governmental organizations have been slow to adopt political marketing tactics (Foster and Lemieux 2012), the use of supporter databases could mitigate their many operational challenges. In this light we must balance concerns about the potential for the nefarious use of technology with the prospects for a favourable effect on political communication and democratic engagement.

The net result of these observations is that Canadians are being offered more targeted, specialized messages by control-obsessed political parties. By segmenting and targeting electors, parties choose who they want to reach and who they decide to ignore. This creates information-rich citizens whose interests are better represented by partisan organizations in their engagements and policies, and information-poor citizens whose needs, demands and aspirations might not be heeded by parties. Political communication then becomes another factor that may contribute to the under-representation or alienation of different categories of citizens deemed less electorally valuable by parties such as opponents' partisans, the economically disadvantaged, Canadian youth, or Aboriginals. In so doing, political institutions seem to reduce interest representation and jeopardize the democratic citizenship of many Canadians.

Conclusion: Political Communication and the State of Canadian Democracy

What are the corresponding implications of the new political communication environment for Canadian democracy? *Political Marketing in Canada* (Marland, Giasson, and Lees-Marshment 2012) reasoned that the democratic implications of marketing strategies and tactics would depend on how Canadian political actors choose to use new tools. On the one hand, technological change has increased the potential for elites to understand and respond to the public's wants and needs, and to communicate efficiently with them. On the other, more sophisticated communication can facilitate attempts to influence media coverage. It can be used to shape the public agenda and to damage opponents' brands. Overall, *Political Marketing in Canada* concluded that political marketing

> can help elevate electors' role in the democratic system, which is a positive development as long as there is still room for politicians to make necessary decisions against public opinion and to explain their reasons for doing so. Overall, there may be a growing sense of a much closer dialogue and connection between citizens and

government in Canada through the use of political marketing, or its use could in fact be entrenching distances. A maturation of political marketing practice means that, in time, political actors should be able to reduce their emphasis on salesmanship and pandering as they move toward more of a deliberative, dialogue-seeking citizen-state relationship. (255)

Many chapters in *Political Communication in Canada* affirm this position. Over time, communicating political messages through innovative technological applications should improve the quality of politics and democracy. However, this is dependent on how political actors choose to employ the tools they have at their disposal. We can see and measure tactics such as advertising, earned media coverage, website content, and tweets, but it is far more difficult to assess the strategy behind these outputs.

We are inclined to affirm Anna Esselment's observation in Chapter 2 that, as a result of these changes, the "political environment that remains is both better and worse." The "plus ça change, plus c'est pareil" Canadian approach to political communication depicted in this book means that to conclude otherwise would be to ignore compelling evidence that the existing habits of political actors are magnifying rather than fundamentally changing. The PMO's digital information subsidies, the parties' use of Twitter, the jostling between political actors and journalists to control the public agenda, the biases in media coverage, the resource constraints faced by interest groups, and the trend toward negative television advertising are among the case studies profiled here that carry a general theme of the persistence of old wine in new bottles. For example, a case could be made that democratic ideals are crumbling because political actors are increasingly insulated from media scrutiny; and yet a contradictory argument could be fashioned that politicians and their critics are now subject to an exponential number of watchdogs with the means to transmit information rapidly. On the whole we must conclude that the quality of democracy is marginally improving because, on balance, the greater the number of free voices participating in the public sphere, the more democratic the society.

Yet communication is in itself an insufficient measure of democracy. It is elections that are considered the defining feature of a democracy as long as they are free, fair, periodic, and competitive (Katz 1997). This includes conditions such as the freedom of expression and association, an independent media, the rule of law, and an autonomous judiciary (Semetko and Scammell 2012). Political communication exists in places that we would consider non-democratic. There are many examples of sham elections, where even though a vote is held, the electoral process is plagued by voter intimidation, banned opposition, vote fraud, or unfair media practices. Rawnsley (2005, 15) points out that dictators, like political leaders in democratic countries, communicate to propagate, mobilize, and legitimate their power. For Sanders (2009), one difference between political communicators in democratic and non-democratic regimes centres on persuasion. In the former, political leaders need words. Words are used to change minds and change behaviours. Political actors make arguments and present evidence in hopes of citizen acquiesces. This process of persuasion is competitive, with different governmental and non-governmental actors vying for the attention of the media and citizens. In the latter, words do not matter; leaders "simply impose views through fear and force" (24). Media characterizations of Canadian leaders as friendly dictators run counter to the prevalence of the rule of law in the Canadian system of government. Specifically, the Charter of Rights and Freedoms provides constitutional protections that include freedom of opinion and expression, freedom of the press and other communication, and freedom of association.

This characterization is therefore a cynical exaggeration perpetuated by government critics. Nevertheless, Canadian leaders unquestionably exercise a level of political control that often appears to be anti-democratic and that merits sombre critical refection. Moreover, declining turnout in elections is a common measure of the health of a democracy, and Canada is one of many countries where the presence of political marketing, negativity, and intense partisanship coexists with fluctuating voter participation.

Additional comparative perspective is warranted to assist readers with arriving at an informed opinion of the implications of the new communication environment on Canadian democracy. Graber, McQuail, and Norris (1998) suggest that democratic communication systems share several key characteristics. First, they feature multiple channels, including radio, television, newspapers, and the Internet, to allow for the rapid distribution of information. Second, the communication process must be open to all, regardless of political interest or the popularity of that interest. The final characteristic of a democratic communication system is a free press. Media scholars identify a number of democratic expectations of the media, which includes the media being a mechanism for surveillance and accountability of the political system, providing meaningful agenda setting, providing a platform for the full range of political actors and ideas, and having independence (Gurevitch and Blumler 1990; Dahlgren 2009). The term "expectation" is key because, as Gurevitch and Blumler (1990) have observed, the media may not always live up to these expectations, and there are democratic implications where this occurs. It is tempting to blame the media for problems with democratic discourse, but it is the decisions by political institutions and their professional media consultants that also contribute to the nature of political communication.

Taking the work of Chapters 2, 4, and 7 together, we can see that Canadian democracy is under strain because of the tensions between the Harper Conservative government and the mainstream news media. These three chapters make it clear that media management strategies make it more difficult for journalists to perform their democratic role. As such, the quality of political information for citizens is at stake, and it is difficult to anticipate the degree to which this will change under different leadership. For, as Chapter 5 explained, there are growing parallels between right-wing and left-wing parties on communications strategies and tactics. Future research should consider how governing parties on different sides of the political spectrum relate to the mass media.

Quality of information is another area where there could be cause for concern. As noted, the nature of a democracy is partly contingent on the quality of information in media content, and in Western democracies there has been a growing focus on the personal lives of politicians (e.g., Stanyer 2007). This personalization is both a function of candidate behaviour and of the news media. But the proper limits of journalistic revelation on private matters such as past romantic partners, sexual orientation, and family members is unclear. This has democratic implications because evaluations of political leaders are an important aspect of voting behaviour, and voters need information about candidates to make judgments (e.g., McAllister 2007). In Chapter 9, Lalancette demonstrated that, for the time being, the focus on the personal lives of politicians is not a prominent feature of Canadian media coverage of leadership races. Even when media stories are personal, they tend to focus on political experience and competence. Although there are certainly outliers – the media's fascination with Justin Trudeau's personal life and family comes to mind – we can observe that the Canadian media treat politicians differently than is the case in other countries. This suggests that Canadians are receiving more useful information about their politicians and that infotainment has not (yet) pervaded the Canadian news media.

Finally, chapters in this book provide some optimism for the role new digital technology can play in encouraging participation in Canadian politics. According to Chadwick (2006, 84), e-democracy involves "efforts to broaden political participation by enabling citizens to connect with one another and with their representatives via new information and communication technologies." At one level, Chapter 6's analysis of party leaders' use of Twitter appears to be inconsistent with Chadwick's definition, given the focus on broadcasting tweets. However, those Canadians who choose to follow a politician on Twitter can receive considerable information about campaign promises and leader views, thereby providing an unmediated source of campaign learning for voters. As well, the optimism surrounding deliberative democracy presents potential conflict with the role of legitimate democratic institutions. For instance, citizens elect a representative to act on their behalf in Parliament, and social media is a tool that can increase legislators' responsiveness to citizens' concerns. But the speeding up of interactive political communication can strengthen party discipline through message discipline, generate public pressure for quick decisions on complex topics, and complicate the ability for elected officials to make unpopular decisions that are for the greater good.

In sum, many chapters in *Political Communication in Canada* speak of an accelerated, information-saturated, strategy-oriented, and increasingly negative context of production and distribution of political messages. Conclusions about what type of citizenry is emerging in this new environment often cast a bleak light and suggest that although there is more political information than ever, many citizens tune out. Other findings offer optimism for the implications of political communication on democratic citizenship as elites relinquish control of mass communication. Canadians have more opportunities to become informed about politics, to engage in democratic debate, and to become active participants in shaping public policy and election outcomes, and yet their participation on Election Day has been declining in the last three decades. It is our impression that, on balance, the political media environment may be speeding up in Canada, but the evolution toward an idealistic form of democratic discourse is glacial.

Glossary

The following concepts are used in *Political Communication in Canada* and/or were key concepts in *Political Marketing in Canada* that provide contextual relevance.

agenda setting: Agenda setting occurs when extensive media coverage of an issue increases people's perceptions of the issue's importance. The more coverage an issue receives, the more likely people are to perceive that issue as the most important issue at the time. As such, agenda setting is an indirect political effect of media coverage on public opinion that concerns the salience of issues within it. The amount of attention devoted to particular issues in the news can influence political priorities and policy responses.

attack ad: Negative advertising that emphasizes the personal characteristics of an opponent, rather than just political or policy aspects. One notorious example in Canada was the Jean Chrétien "face" ads featured during the 1993 federal election campaign.

blog: A blog (short for weblog) is a low-cost, publicly available single or multi-authored web publication channel that has limited external editorial oversight. It provides updated mixed-media information, commentary, and opinions that are archived in reverse chronological order and regularly comprises interactive elements such as hyperlinks that redirect audience members to other digital material. Blog writers (bloggers) post content on a wide variety of themes and subjects, including politics. The aggregation of all blogs is referred to in the literature as the blogosphere. See also *social media*.

branding: Less tangible and specific than a product or organization, branding is the overall perception of it, which often employs familiar logos or slogans to evoke meanings, ideas, and associations in the consumer. In politics, branding involves creating a trustful long-term relationship with electors. A political brand is a deeper construct than a political image because brands integrate personal experiences, emotional attachments, and partisan loyalties. Political parties tend to renew their brands when a new party leader is installed. One example of a major Canadian political rebranding effort occurred after the late 2003 merger of the Canadian Alliance party with the federal Progressive Conservative Party to become the "new" Conservative Party. See also *image*.

Canadian Election Study (CES): A publicly available dataset of Canadian attitudes and opinions. Survey data are collected at each federal general election.

centralization: This concept posits that information, power, and communication strategy are clustered among a nucleus of core decision makers. In the federal government, centralization normally refers to central agencies, such as the Prime Minister's Office (PMO), Privy Council Office (PCO), Treasury Board, and Department of Finance. Within political parties, it normally refers to a leader's concentration of power and their inner circle.

database marketing: The collection of quantitative data about electors that are stored in a database and used to create targeted marketing messages. In Canada, political parties begin with information obtained from the list of electors, to which they add information collected when people contact the party, take out a membership, make a monetary donation, take a lawn sign, respond to GOTV contact efforts, and other data sources. More sophisticated operations may obtain information from Statistics Canada, telemarketing companies, and list providers. Increasingly, database marketing is supplemented by scouring the web and social media for information about electors. List management software programs used by Canadian political parties include Liberalist, NDP Vote, and the Conservative Party's Constituency Information Management System (CIMS). See also *direct marketing; micro-targeting; relationship marketing.*

direct marketing: The communication of precise messages – via direct mail, telemarketing, direct dialogues, personalized emails, or texts to portable communication devices – directly to individuals, thereby bypassing other filters such as the mass media. Direct marketing relies on lists of consumers generated by companies that collect data on individuals concerning a range of factors, such as geographical location, age, and lifestyle, which are often used in conjunction with segmentation. In politics, informal lists have conventionally been generated by the results of door-to-door canvassing or phone calls that record voting intentions, but the process is becoming more complex. Political organizations often purchase commercial lists, or use databases, to build up voter profiles.

election platform/manifesto: A document identifying a political party's commitments and policy proposals that, should the party form government, will be used to guide the government's agenda. A notable Canadian example is the Liberal Red Book.

election turnout: The proportion of electors who voted in an election. Voter turnout in Canada has been decreasing over time.

frames: Frames are organized patterns of interpretive cues used to present and give meaning to social and political issues by emphasizing and excluding specific elements. The political elite, as well as the media, develop frames to define political debate. Frames provide targeted audiences with definitions of social and political problems or issues, along with the agendas of political actors who are involved in the debate and their proposed solutions or potential outcomes. See also *framing; image.*

framing: Framing is both a strategic communicative process and an effect of the mediatization of politics. As a strategic process, it is the act of shaping or presenting issues, such as political ones, by using frames in order to reflect particular agendas and influence or sway public opinion. As an effect, framing is the indirect consequence of exposure to media coverage of politics over public opinion. The selection of frames used by the media to depict a political issue brings viewers to cast differing forms of responsibility on policy makers or to interpret political and social issues in varying ways. The public's understanding of political issues is therefore influenced according to the dominant frame used in the press to define such issues. See also *frames; image.*

gatekeeping: An information selection process in which news editors select and favour certain types of stories over others, thereby controlling the flow and content of information, political or otherwise, and ultimately determining the content of the news. A professional journalistic routine shown to be highly influenced by the subjective attitudes or biases of journalists and editors and resulting in a hierarchy among news stories presented to the public at the expense of the public interest. See also *agenda setting.*

get out the vote (GOTV): Mobilization strategies and tactics designed to ensure that supporters will turn out to cast a ballot on election day. GOTV increasingly uses segmentation and voter profiling to identify who to target, and direct marketing (for example emails, mobile texts, or phone calls) is often employed to reach the key targeted segments.

horserace journalism: The tendency of news media to report predominantly on opinion polls, campaign events, or an election campaign leader at the expense of electoral issues. This results in the media coverage of even routine political events becoming a mini-contest in which the focus is on winning and losing. As an election campaign climaxes, there is less coverage on contestants who are behind in opinion polls and thus cannot influence the outcome of the race to the finish line. See also *gatekeeping.*

image: The mental impression or perceptions of an object, being, or concept, such as a politician, political party, or a public policy. Images are subjective in nature due to the varying ways in which their target audiences receive, absorb, process, and evoke political communication. The public images of political actors are actually imaginary constructs shaped by information and visuals that are controlled and filtered by political parties, public relations personnel, the media, pundits, and others. See also *branding; framing.*

information subsidies: Newsworthy information packaged in a manner that is designed to be easily reproduced by the media, such as a press conference. Increasingly, this means low-cost information that is distributed electronically by political organizations, such as news releases, social media posts, blogs, tweets, digital photographs, and online video. Newsrooms benefit from speedy information with little effort and audiences have access to a wider variety of news. However, the objectivity and integrity of such controlled content is suspect, especially if the source has political motives.

inter-election period: The time between official election campaign periods. In Canada, the inter-election period, unlike campaign periods, is not regulated by extraordinary limitations on fundraising activities, spending, or political communications.

Internet marketing: The use of digital technologies, including the web, email, and mobile devices to achieve marketing objectives. Electronic communication platforms and tactics include websites, wikis, and e-newsletters and texting, social networking, online file sharing, social bookmarking, and micro-blogging. In political marketing, this practice is sometimes referred to as e-marketing, but it involves more than just online communication. Instead, it incorporates concepts such as market orientation and relationship marketing in order to tailor the message to the receiver, enable two-way communication between political elites and the public, and build a long-term relationship.

market intelligence: Empirical data about the political marketplace and public views, also known as "market research." The practice of collecting market intelligence involves the use of quantitative and qualitative methods such as polls, opinion surveys, focus groups, role playing, co-creation, and consultation, as well as analysis of existing public census data and election records. Politicians and political parties rely on market intelligence to aid in prioritizing issues, developing communication strategies, and helping to present themselves as the most competent alternative able to address those issues.

market-oriented party: A political party that uses market intelligence to assist in the identification and understanding of electors' concerns and priorities and incorporates them into the design of the party's product offerings before finalizing political decisions. A market-oriented party therefore engages in far more consultation and dialogue with the electorate than does a product-oriented party or sales-oriented party. See also *product-oriented party; sales-oriented party.*

mass media or news media: Print and broadcast news outlets that reach a mass audience, such as newspapers, magazines, radio, and television. They are also referred to as the mainstream/traditional/conventional media due to the growing presence of alternative information channels including blogs, online media, community news outlets, transit publications, social media, and other non-traditional news sources.

message event proposal (MEP): A policy instrument created by communications personnel in Stephen Harper's PMO that requires departments to provide media information to the Privy Council Office and the Prime Minister's Office about their plans for public events, such as ministerial announcements. MEPs identify details of the planned event, the spokesperson, attendees, desired media headlines, the intended audiences, key messages, photos, and attire. This strategy allows the centre to ensure that ministers' messages are consistent in style, tone, and substance with those of the prime minister himself. See also *centralization.*

micro-targeting: A strategic use of resources, uncovered through market intelligence, designed to focus communication efforts on small segments of the electorate whose socio- and geo-demographic profiles indicate a propensity for supporting the sponsor. Sometimes called hyper segmentation, this process relies on complex voter profiling activities or databases. One such example in Canada was the Conservatives' use of micro-policies such as boutique tax credits targeted to construction workers and truck drivers.

narrowcasting: The act of selecting media, based on the nature of the communication, which are most likely to reach targeted market segments. For example, communicating with target groups by advertising on sports or lifestyle specialty channels instead of via the broader mass media.

non-profit organizations: Non-profit organizations (also known as nonprofits) comprise a broad category of political actors such as special interest groups, social movement organizations, and service and charity organizations that publicly advocate in favour of causes and issues in social and political debates. These organizations share many commonalities such as a formal organizational structure, operations separate from the government and business sectors, the reinvestment of earnings back into the organization, self-governing and voluntary practices, and the desire to achieve a collectively determined normative goal or mission.

partisanship: A person's psychological ties to a political party. Every party has a core of strong partisans who may or may not publicly self-identify as such.

permanent campaign: Sustained electioneering after the conclusion of an election campaign and throughout the inter-election period. This constant campaigning increases in intensity as the next election approaches. In the Canadian parliamentary system of government, this is more prevalent during a period of minority government because of the proximity to, and uncertainty of, the next election campaign.

personalization: The self-disclosure of private and personal details by politicians and the increased attention from the news media on the private life of politicians.

political advertising: Any controlled, mediated (print, television, radio, and Internet), and paid forms of communication created by political actors (governments, political parties, candidates, interest groups) whose objective is to influence the opinions, choice, or behaviour of their destined audiences. See also *attack ad.*

political communication: The role of communication in politics including the generation of messages by political actors (political organizations, non-profits, citizens, the media) and their transmission as well as reception. It occurs in a variety of forms (formal or informal) and venues (public and private) and through a variety of media (mediated or unmediated content).

political marketing: The application of business-marketing concepts to the practice and study of politics and government. With political marketing, a political organization uses business techniques to inform and shape its strategic behaviours that are designed to satisfy citizens' needs and wants. Strategies and tools include branding, e-marketing, delivery, focus groups, GOTV, internal marketing, listening exercises, opposition research, polling, public relations, segmentation, strategic product development, volunteer management, voter-driven communication, voter expectation management, and voter profiling.

press gallery: An organization composed of journalists accredited to cover the activities of the legislature. For example, the Ottawa-based Canadian Parliamentary Press Gallery, Quebec's Tribune de la presse at the National Assembly, or the White House Correspondents' Association in the United States.

product-oriented party: A party that employs a marketing strategy guided by the assumption that electors will recognize the normative value of the party's ideas and will thus vote for it. Little consideration is given to gathering and using market intelligence to design or communicate the party's product offering. See also *market-oriented party; sales-oriented party.*

public opinion research (POR): The collection of intelligence from a sample of the population that is designed to measure the public's views on issues, policies, leaders, and parties. The most common forms of POR are opinion surveys and focus groups, which can be purchased on a customized or omnibus basis.

public relations: The strategic use of communication tools and media relations techniques to optimize interactions between an organization and its stakeholders.

relationship marketing: The use of marketing to build customer relationships and long-term associations that are sustained through commitment, loyalty, mutual benefit, and trust. See also *direct marketing.*

sales-oriented party: A party that uses market intelligence to design strategies for selling its products to targeted segments of the electorate. Emphasis is placed on research for advertising and message design as opposed to the design of the party's actual product offering. See also *market-oriented party; product-oriented party.*

segmentation: Division of electors into new groups to allow more efficient targeting of political resources, and also creation of new segments, such as ethnic minorities or seniors, as society evolves. Segments can be targeted by policy, communication, or GOTV, as well as to encourage greater volunteer activity. See also *micro-targeting.*

social media: Internet-based applications characterized by the ability to enable users to create and share content and collaborate and communicate with users. Includes applications such as social networking (Facebook), blogs, microblogs (Twitter), online video (YouTube), wikis (Wikipedia), and social bookmarking (Digg), which are also known as Web 2.0.

Partial source: *Political Marketing in Canada* (Vancouver: UBC Press, 2012): 257-63.

References

Abele, Andrea. 1985. Thinking about thinking: Causal, evaluative, and finalistic cognitions about social situations. *European Journal of Social Psychology* 15(3): 315-32.

Abélès, Marc. 2007. *Le Spectacle du Pouvoir*. Paris: L'Herne.

Abu-Laban, Yasmeen, and Linda Trimble. 2010. Covering Muslim Canadians and politics in Canada: The print media and the 2000, 2004 and 2006 federal elections. In S. Sampert and L. Trimble, eds., *Mediating Canadian Politics*, 129-50. Scarborough, ON: Pearson Education.

Adams, Christopher. 2010. Polling in Canada: Calling the election. In S. Sampert and L. Trimble, eds., *Mediating Canadian Politics*, 151-68. Scarborough, ON: Pearson Education.

Adatto, Kiku. 2008. *Picture Perfect: Life in the Age of the Photo Op*. Princeton, NJ: Princeton University Press.

Adriaansen, Maud L., Philip van Praag, and Claes H. de Vreese. 2010. Substance matters: How news content can reduce political cynicism. *International Journal of Public Opinion Research* 22(4): 433-57.

Advertising Standards Canada. 2012. The Canadian Code of Advertising Standards. http://www.adstandards.com/.

Akin, David. 2009. Olympic logo called too conservative. *Windsor Star*, October 2, A1.

Alboim, Elly. 2012. On the verge of total dysfunction: Government, media, and communications. In D. Taras and C. Waddell, eds., *How Canadians Communicate IV: Media and Politics*, 45-53. Edmonton: Athabasca University Press.

Aldrich, John H., and David W. Rohde. 2000. The logic of conditional party government: Revisiting the electoral connection. In L.C. Dodd and B.I. Oppenheimer, eds., *Congress Reconsidered*, 7th ed., 249-70. Washington, DC: CQ Press.

Allan, Stuart. 2006. *Online News: Journalism and the Internet*. Berkshire, UK: Open University Press.

Allen, Barbara, and Daniel Stevens. 2010. Truth in advertising? Visuals, sound, and the factual accuracy of political advertising. Paper presented at the annual meeting of the American Political Science Association, Seattle, September.

Altheide, David L. 1996. *Qualitative Media Analysis*. Thousand Oaks, CA: Sage.

American National Election Study. 2008. Strength of partisanship 1952-2008. *ANES Guide to Public Opinion and Electoral Behavior*. http://www.electionstudies.org/.

Ancu, Monica. 2011. From soundbite to textbite: Election 2008 comments on Twitter. In J.A. Hendricks and L.L. Kaid, eds., *Techno Politics in Presidential Campaigning: New Voices, New Technologies, and New Voters*, 11-21. New York: Routledge.

Anderson, Terry, and Heather Kanuka. 2003. *E-research: Methods, Strategies and Issues.* Boston: Pearson Education.

Andrew, Blake. 2007. Media-generated shortcuts: Do newspaper headlines present another roadblock for low-information rationality? *Harvard International Journal of Press/Politics* 12(2): 24-43.

Andrew, Blake, Patrick Fournier, and Stuart Soroka. 2013. The Canadian party system: Trends in election campaign reporting, 1980-2008. In A. Bittner and R. Koop, eds., *Parties, Elections, and the Future of Canadian Politics*, 161-84. Vancouver: UBC Press.

Androich, Alicia. 2011. Canada's newspaper circulation numbers still dropping. *Marketing Magazine*, November 1. http://www.marketingmag.ca/.

Ankersmit, Frank. 2002. *Political Representation.* Stanford, CA: Stanford University Press.

Ansolabehere, Stephen, and Shanto Iyengar. 1994. Of horseshoes and horse races: Experimental studies of the impact of poll results on electoral behavior. *Political Communication* 11(4): 413-30.

–. 1995. *Going Negative: How Political Advertisements Shrink and Polarize the Electorate.* New York: Free Press.

Areni, Charles S., Elizabeth M. Ferrell, and James B. Wilcox. 2000. The persuasive impact of reported group opinions on individuals low vs. high in need for cognition: Rationalization vs. biased elaboration? *Psychology and Marketing* 17(10): 855-75.

Aucoin, Peter. 2008. New public management and new public governance: Finding the balance. In D. Siegel and K. Rasmussen, eds., *Professionalism and Public Service: Essays in Honour of Kenneth Kernaghan*, 16-33. Toronto: University of Toronto Press.

Axsom, Danny, Suzanne Yates, and Shelly Chaiken. 1987. Audience response as a heuristic cue in persuasion. *Journal of Personality and Social Psychology* 53(1): 30-40.

Baarda, Rachel, and Rocci Luppicini. 2012. The use and abuse of digital democracy: Case study of Mybarackobama.com. *International Journal of Technoethics* 3(3): 50-68.

Babbie, Earl. R. 1998. *The Practice of Social Science.* Belmont, CA: Wadsworth.

Bailey, Michael, Lee Sigelman, and Clyde Wilcox. 2003. Presidential persuasion on social issues: A two-way street? *Political Research Quarterly* 56(1): 49-58.

Baker, Sara, and Richard Petty. 1994. Majority and minority influence: Source-position imbalance as a determinant of message scrutiny. *Journal of Personality and Social Psychology* 67(1): 5-19.

Balagus, Michael. 2011. Former premier's chief of staff and campaign director, Manitoba NDP. Personal communication, August 15.

Baluja, Tamara. 2012. 7 interesting facts from the newspaper Canada 2012 circulation data report. *The Canadian Journalism Project.* April 9. http://j-source.ca/.

Banting, Keith, Stuart Soroka, and Edward Koning. 2013. Multicultural diversity and redistribution. In K. Banting and J. Myles, eds., *Inequality and the Fading of Redistributive Politics*, 165-86. Vancouver: UBC Press.

Bardoel, Jo. 1996. Beyond journalism: A profession between information society and civil society. *European Journal of Communication* 11(3): 283-302.

Barney, Darin. 2005. *Communication Technology*. Vancouver: UBC Press.

Bartels, Larry. 2003. Democracy with attitudes. In M. MacKuen and G. Rabinowitz, eds., *Electoral Democracy*, 48-82. Ann Arbor: University of Michigan Press.

Bashevkin, Sylvia. 2009a. "Stage" versus "actor" barriers to women's federal party leadership. In S. Bashevkin, ed., *Opening Doors Wider: Women's Political Engagement in Canada*, 108-26. Vancouver: UBC Press.

–. 2009b. *Women, Power, Politics: The Hidden Story of Canada's Unfinished Democracy*. Don Mills, ON: Oxford University Press.

Bastedo, Heather, Wayne Chu, and Jane Hilderman. 2012. Occupiers and legislators: A snapshot of political media coverage. *Samara Democracy Reports*. http://www.samaracanada.com/.

Bastedo, Heather, Wayne Chu, Jane Hilderman, and André Turcotte. 2011. The real outsiders: Politically disengaged views on politics and democracy. *Samara Democracy Reports*. http://www.samaracanada.com/.

Baum, Matthew A. 2007. Soft news and foreign policy: How expanding the audience changes the policies. *Japanese Journal of Political Science* 8(1): 115-45.

Baumgartner, Frank R., and Bryan D. Jones. 1993. *Agendas and Instability in American Politics*. Chicago: University of Chicago Press.

–. 1994. Attention, boundary effects, and large-scale policy change in air transportation policy. In D.A. Rochefort and R.W. Cobb, eds., *The Politics of Problem Definition: Shaping the Policy Agenda*, 50-66. Lawrence: University Press of Kansas.

Bélanger, Éric. 2003. Issue ownership by Canadian political parties, 1953-2001. *Canadian Journal of Political Science* 36(3): 539-58.

Bélanger, Éric, and Bonnie Meguid. 2008. Issue salience, issue ownership, and issue-based vote choice. *Electoral Studies* 27(3): 477-91.

Bennett, Colin J., and Robin M. Bayley. 2012. Canadian federal political parties and personal privacy protection: A comparative analysis. Report prepared for the Office of the Privacy Commissioner of Canada. http://www.priv.gc.ca/.

Bennett, W. Lance. 1998. The uncivic culture: Communication, identity and the rise of lifestyle politics. *PS: Political Science and Politics* 31(4): 740-61.

–. 2004. Gatekeeping and press-government relations: A multigated model of news construction. In L.L. Kaid, ed., *Handbook of Political Communication Research*, 283-313. Mahwah, NJ: Lawrence Erlbaum.

Bennett, W. Lance, and Shanto Iyengar. 2008. A new era of minimal effects? The changing foundations of political communication. *Journal of Communication* 58(4): 707-31.

Bennett, W. Lance, and Jarol B. Manheim. 2001. The big spin: Strategic communication and the transformation of pluralist democracy. In L.W. Bennett and R.M. Entman, eds., *Mediated Politics: Communication and the Future of Democracy*, 279-98. Cambridge: Cambridge University Press.

Benoit, Liane E. 2006. Ministerial staff: The life and times of Parliament's statutory orphans. In *Canada, Commission of Inquiry into the Sponsorship Program and Advertising Activities: Restoring Accountability*, vol. 1, 145-252. Ottawa: Public Works and Government Services Canada.

Bentivegna, Sara. 2002. Politics and the new media. In L.A. Lievrouw and S.M. Livingstone, eds., *The Handbook of New Media*, 50-61. London: Sage.

Berkowitz, D., and Douglas B. Adams. 1990. Information subsidy and agenda-building in local television news. *Journalism Quarterly* 67(4): 723-31.

Berry, Jeffrey. 2003. *A Voice for Non-Profits*. Washington, DC: Brookings Institution Press.

Berthiaume, Lee. 2011. Conservatives admit they're behind false byelection phone calls in Liberal riding. *National Post*, November 30. http://news.nationalpost.com/.

Bertot, John Carlo, Paul T. Jaeger, Sean Munson, and Tom Glaisyer. 2010. Social media technology and government transparency. *Computer* 43(11): 53-59.

Bimber, Bruce. 1999. The Internet and citizen communication with government: Does the medium matter? *Political Communication* 16(4): 409-28.

–. 2001. Information and political engagement in America: The search for effects of information technology at the individual level. *Political Research Quarterly* 54(1): 53-67.

–. 2003. *Information and American Democracy: Technology in the Evolution of Political Power*. Cambridge: Cambridge University Press.

Bittner, Amanda. 2011. *Platform or Personality? The Role of Party Leaders in Elections*. Oxford: Oxford University Press.

Blais, André, and Martin M. Boyer. 1996. Assessing the impact of televised debates: The case of the 1988 Canadian election. *British Journal of Political Science* 26(2): 143-64.

Blais, André, and Agnieszka Dobrzynska. 2003. Life cycle, generational, and period effects in turnout. Paper presented at the 2003 Joint Sessions of the European Consortium for Political Research, Edinburgh, March-April.

Blais, André, Elisabeth Gidengil, Richard Nadeau, and Neil Nevitte. 2001. Measuring party identification: Canada, Britain, and the United States. *Political Behavior* 23(1): 5-22.

–. 2002. *Anatomy of a Liberal Victory: Making Sense of the Vote in the 2000 Canadian Election*. Peterborough, ON: Broadview Press.

Blick, Andrew. 2004. *People Who Live in the Dark: The History of the Special Advisor in British Politics*. London: Politico's.

Blumenthal, Sidney. 1980. *The Permanent Campaign: Inside the World of Elite Political Operatives*. Boston: Beacon Press.

Blumer, Herbert. 1946. Collective behaviour. In A.M. Lee, ed., *New Outlines of the Principles of Sociology*, 166-222. New York: Barnes and Noble.

Blumler, Jay G., and Michael Gurevitch. 1995. *The Crisis of Public Communication*. London: Routledge.

Blumler, Jay G., and Dennis Kavanagh. 1999. The third age of political communication: Influences and features. *Political Communication* 16(3): 209-30.

Boesveld, Sarah. 2013. #dayinthelife: Stephen Harper tweets his entire day, including breakfast with his cat. *National Post*, January 28. http://news.nationalpost.com/.

Bohan, Caren. 2012. President of the White House Correspondents' Association, 2011-12. Personal communication, April 3.

Boivin, Simon. 2005. Course à la direction du Parti Québécois: Le clan Marois peaufine son offensive. *Le Soleil,* July 23, A16.

Bonk, Kathy, Henry Griggs, and Emily Tynes. 1999. *The Jossey-Bass Guide to Strategic Communication for Non-Profits.* San Francisco: Jossey-Bass.

Boorstin, Daniel J. 1992. *The Image: A Guide to Pseudo-Events in America.* New York: Vintage Books.

Borge, Rosa, and Ana Sofia Cardenal. 2010. Surfing the Net: A pathway to political participation without motivation? Paper presented at the Internet, Policy, and Politics 2010: An Impact Assessment conference, Oxford, September.

Bowers, Chris, and Matthew Stoller. 2005. *Emergence of the progressive blogosphere: A new force in American politics.* New Politics Institute. http://ndn-newpol.civic actions.net/.

Bowman, Karyn. 2000. Polling to campaign and to govern. In N. Ornstein and T. Mann, eds., *The Permanent Campaign and Its Future,* 54-74. Washington, DC: American Enterprise Institute and Brookings Institution.

Boyd, Danah, Scott Golder, and Gilad Lotan. 2010. Tweet, tweet, retweet: Conversational aspects of retweeting on Twitter. Paper presented at the Hawaii International Conference on System Sciences, Kauai, Hawaii, January.

Braaten, Aaron. 2005. *The Great Canadian Blog Survey: A Snapshot of the Canadian Blogosphere in 2005.* Edmonton: University of Alberta Press.

Brader, Ted. 2005. Striking a responsive chord: How political ads motivate and persuade voters by appealing to emotions. *American Journal of Political Science* 49(2): 388-405.

–. 2006. *Campaigning for Hearts and Minds: How Emotional Appeals in Political Ads Work.* Chicago: University of Chicago Press.

Bradley, Margaret M., and Peter J. Lang. 1999. Fearfulness and affective evaluations of pictures. *Motivation and Emotion* 23(1): 1-13.

Bradley, Samuel D., James R. Angelini, and Sungkyoung Lee. 2007. Psychophysiological and memory effects of negative political ads. *Journal of Advertising* 36(4): 115-27.

Brants, Kees, and Katrin Voltmer. 2011. Introduction: Mediatization and de-centralization of political communication. In K. Brants and K. Voltmer, eds., *Political Communication in Postmodern Democracy: Challenging the Primacy of Politics,* 1-16. New York: Palgrave Macmillan.

Brennan, Richard, Tonda MacCharles, and Joanna Smith. 2008. Harper a man who "lives in a bubble." *Toronto Star,* September 25. http://www.thestar.com/.

Brosius, Hans-Bernd, and Hans M. Kepplinger. 1992. Linear and nonlinear models of agenda-setting in television. *Journal of Broadcasting and Electronic Media* 36(1): 5-23.

Budge, Ian, and Dennis J. Farlie. 1983. *Explaining and Predicting Elections.* London: Allen and Unwin.

Bumstead, Justin. 2012. Mapping the Alberta election 2012. Blog. http://www.justin bumstead.ca/.

Burton, Michael J., and Daniel M. Shea. 2010. *Campaign Craft: The Strategies, Tactics, and Art of Political Campaign Management.* Santa Barbara: Praeger.

Cacioppo, John. T., and Richard E. Petty. 1982. The need for cognition. *Journal of Personality and Social Psychology* 42(1): 116-30.

Cacioppo, John T., Richard E. Petty, Jeffrey A. Feinstein, and Blair W.G. Jarvis. 1996. Dispositional differences in cognitive motivation: The life and times of individuals varying in need for cognition. *Psychological Bulletin* 119(2): 197-253.

Campion-Smith, Bruce. 2008. Auditor balks at vetting by PMO. *Toronto Star,* May 1. http://www.thestar.com/.

Campus, Donatella. 2010. Mediatization and personalization of politics in Italy and France: The cases of Berlusconi and Sarkozy. *International Journal of Press/ Politics* 15(2): 219-35.

Canada. 2011. *Canada Year Book, 2011.* http://www.statcan.gc.ca/.

–. 2012a. Constitutional documents. Department of Justice, Justice Laws. http://laws-lois.justice.gc.ca/.

–. 2012b. *2010-11 Annual Report on Government of Canada Advertising Activities. Public Works and Government Services Canada.* http://www.tpsgc-pwgsc.gc.ca/.

–. 2014. Government electronic directory services. http://sage-geds.tpsgc-pwgsc.gc.ca/.

Canada Revenue Agency. 2003. Political activities: Policy statement. Ottawa, ON. http://www.cra-arc.gc.ca.

Canadian Association of Journalists. 2010. An open letter to Canadian journalists. CAJ, June. http://www.caj.ca/.

Canadian Election Study. 2011. Canadian Opinion Research Archive. http://ces-eec.org/.

Canadian Press. 2007. Harper's Jewish mailing list draws privacy inquiry. CTV News, October 11. http://www.ctvnews.ca/.

–. 2011a. Bureaucrats told to "Harperize" government message. CBC News, November 29. http://www.cbc.ca/.

–. 2011b. Harper government's ad buy costs taxpayers $26m. CTV News, March 13. http://www.ctvnews.ca/.

–. 2011c. Social media takes off in first week of campaign. News 1130, April 2. http://www.news1130.com.

–. 2012. Former Tory MP had misgivings about voter ID system. CBC News, March 16. http://www.cbc.ca/.

Capara, Gian Vittorio, Michele Vecchione, Claudio Barbaranelli, and Chris R. Fraley. 2007. When likeness goes with liking: The case of political preference. *Political Psychology* 28(5): 609-32.

Cappella, Joseph N., and Kathleen Hall Jamieson. 1997. *Spiral of Cynicism: The Press and the Public Good.* Oxford: Oxford University Press.

Carmines, Edward, and James Stimson. 1980. The two faces of issue voting. *American Political Science Review* 74(1): 78-91.

Carpentier, Nico. 2011. New configurations of the audience? The challenges of user-generated content for audience theory and media participation. In V. Nightingale, ed., *The Handbook of Media Audiences,* 190-212. Malden, MA: Wiley-Blackwell.

Carstairs, Catherine. 2006. Roots nationalism: Branding English Canada cool in the 1980s and 1990s. *Social History* 39(77): 235-55.

Carty, R. Kenneth, and William Cross. 2010. Political parties and the practice of broker-age politics. In J.C. Courtney and D.E. Smith, eds., *The Oxford Handbook of Canadian Politics,* 191-207. Oxford: Oxford University Press.

Carty, R. Kenneth, William Cross, and Lisa Young. 2000. *Rebuilding Canadian Party Politics.* Vancouver: UBC Press.

CBC News. 2006. Harper says he is finished with Ottawa press corps. May 24. http://www.cbc.ca/.

–. 2008. Tories use RCMP to block media from talking to candidate Cadman. September 23. http://www.cbc.ca/.

– 2009. Tory logos on federal cheques draw fire. October 14. http://www.cbc.ca/.

–. 2012a. Air Canada brings in senior Harper aide Derek Vanstone. July 12. http://www.cbc.ca/.

–. 2012b. Privy Council spending $463K on ethnic media monitoring. November 15. http://www.cbc.ca/.

CBC Radio-Canada. 2013. Who we are, what we do. http://cbc.radio-canada.ca/.

CEFIO (Centre Francophone d'Informatisation des Organisations). 2010. L'explosion des médias sociaux au Québec. *NETendances 2010* 1(1): 1-18.

Cha, Meeyoung, Haddadi Hamed, Fabrício Benevenuto, and Krishna P. Gummadi. 2010. Measuring user influence in Twitter: The million follower fallacy. In M. Hearst, W. Cohen, and S. Gosling, eds., *Proceedings of the Fourth International Association for the Advancement of Artificial Intelligence Conference on Weblogs and Social Media,* 10-17. Washington, DC: AAAI Press.

Chadwick, Andrew. 2006. *Internet Politics: States, Citizens, and New Communication Technologies.* London: Oxford University Press.

Chaffee, Steven H., ed. 1975. *Political Communication: Issues and Strategies for Research.* Beverly Hills: Sage.

Chase, Steven, Tamara Baluja, and Jane Taber. 2011. The bubble: Conservative tactics keep campaign off Main Street, on script. *Globe and Mail,* April 5. http://www.theglobeandmail.com/.

Cheadle, Bruce. 2011. Harper's economic action plan website got approval despite violat-ing rules. *Whitehorse Star,* January 6, 8.

Cheadle, Bruce, and Stephanie Levitz. 2012. Government paid for media monitoring of immigration minister's image. *Globe and Mail,* November 13. http://www.theglobeandmail.com/news/.

Cheng, Hong, and Daniel Riffe. 2008. Attention, perception, and perceived effects: Negative political advertising in a battleground state of the 2004 presidential elec-tion. *Mass Communication and Society* 11(2): 177-96.

Chong, Dennis, and James N. Druckman. 2007. Framing theory. *Annual Review of Political Science* 10: 103-26.

Chu, Wayne. 2007. "Of blogs and broadcasters": The influence of Web logs in electoral campaigns. Paper presented at the annual meeting of the Canadian Political Science Association, Saskatoon, May.

Clark, Campbell. 2006. Harper restricts ministers' message. *Globe and Mail,* March 17, A1.

Clark, Campbell, Gloria Galloway, Steven Chase, Daniel Leblanc, Jane Taber, and Bill Curry. 2010. The year that was in Canadian politics. *Globe and Mail,* December 27. http://www.theglobeandmail.com/.

Clarke, Harold D., Thomas J. Scotto, and Allan Kornberg. 2011. Valence politics and economic crisis: Electoral choice in Canada 2008. *Electoral Studies* 30(3): 438-49.

CMRC (Canadian Media Research Consortium). 2011a. Canadian consumers unwilling to pay for news online. http://www.cmrcccrm.ca/.

–. 2011b. Canadians value home Internet connection more than other media devices. http://www.cmrcccrm.ca/.

–. 2011c. Even in the digital era, Canadians have confidence in mainstream news media. http://www.cmrcccrm.ca/.

–. 2011d. Social networks transforming how Canadians get the news. http://www.cmrcccrm.ca/.

Cobb, Roger W., and Charles D. Elder. 1983. *Participation in American Politics: The Dynamics of Agenda-Building.* Boston: Allyn and Bacon.

Cohen, Bernard C. 1963. *The Press and Foreign Policy.* Princeton, NJ: Princeton University Press.

Coleman, Stephan, and Jay G. Blumler. 2009. *The Internet and Democratic Citizenship.* Cambridge: Cambridge University Press.

ComputerWorld. 2011. Canada's first social media election missed the mark. *IT World Canada,* May 13. http://www.itworldcanada.com/.

Conaghan, Catherine, and Carlos de la Torre. 2008. The permanent campaign of Rafael Correa: Making Ecuador's plebiscitary presidency. *International Journal of Press/ Politics* 13(3): 267-84.

Conservative Party of Canada. n.d. (*c.* 2007). CIMS and your campaign. Internal party document. http://thestar.blogs.com/files/cims.ppt.

–. 2011a. Rising to the challenge. http://www.youtube.com/.

–. 2011b. 10 things you might not know about Prime Minister Stephen Harper. http://www.conservative.ca/.

–. 2013. CIMS. http://cpccims.ca/.

Cook, Corey. 2002. The permanence of the "permanent campaign": George W. Bush's public presidency. *Presidential Studies Quarterly* 32(4): 753-64.

Cormack, Patricia. 2012. Double-double: Branding, Tim Hortons, and the public sphere. In A. Marland, T. Giasson, and J. Lees-Marshment, eds., *Political Marketing in Canada,* 209-23. Vancouver: UBC Press.

Corner, John. 2003. Mediated persona and political culture. In J. Corner and D. Pels, eds., *Media and the Restyling of Politics: Consumerism, Celebrity and Cynicism,* 67-84. London: Sage.

Corner, John, and Dick Pels. 2003. Introduction: The re-styling of politics. In J. Corner and D. Pels, eds., *Media and the Restyling of Politics: Consumerism, Celebrity and Cynicism,* 1-17. London: Sage.

Cosgrove, Kenneth M. 2007. *Branded Conservatives: How the Brand Brought the Right from the Fringes to the Center of American Politics.* New York: Peter Lang.

CPPG (Canadian Parliamentary Press Gallery). 1987. Constitution. CPPG. http://press-presse.parl.gc.ca/.

–. 2006a. Minutes of the annual general meeting of the Canadian Parliamentary Press Gallery. April 6, Ottawa.

–. 2006b. Minutes of the special general meeting of the Canadian Parliamentary Press Gallery Concerning Relations with the Prime Minister's Office. October 23, Ottawa.

–. 2012. Minutes of the Executive Committee of the Canadian Parliamentary Press Gallery. May 9, Ottawa.

–. 2013. Members list by organization. http://press-presse.parl.gc.ca/.

–. n.d. Handbook. http://press-presse.parl.gc.ca/.

Cronkite, Walter. 1984. Free press is democracy. In *Leading Journalists Tell What a Free Press Means to America*. Chicago: The Society.

Cross, William. 2004. *Political Parties*. Vancouver: UBC Press.

Cross, William, and Lisa Young. 2004. The contours of political party membership in Canada. *Party Politics* 10(4): 427-44.

Cruz, Veronica. 2001. Advocacy in non-profit human service organizations. PhD diss., University of Albany, NY.

CTV News. 2013. Mulcair apologizes for running stop signs. http://www.ctvnews.ca/politics/.

Cunningham, Stanley B. 1999. The theory and use of political advertising. In W.I. Romanow, M. de Repentigny, S.B. Cunningham, W.C. Soderlund, and K. Hildebrand, eds., *Television Advertising in Canadian Elections: The Attack Mode*, 11-25. Waterloo, ON: Wilfrid Laurier University Press.

Curtin, Patricia A. 1999. Reevaluating public relations information subsidies: Market-driven journalism and agenda-building theory and practice. *Journal of Public Relations Research* 11(1): 53-90.

Dahlgren, Peter. 2009. *Media and Political Engagement: Citizens, Communication, and Democracy*. Cambridge: Cambridge University Press.

Dakhlia, Jamil. 2009. Du populaire au populisme? Idéologie et négociation des valeurs dans la presse *people* française. *Communication* 27(1): 66-82.

–. 2010. *Mythologie de la Peopolisation*. Paris: Le Cavalier bleu.

Dardis, Frank E., Fuyuan Shen, and Heidi H. Edwards. 2008. Effects of negative political advertising on individuals' cynicism and self-efficacy: The impact of ad type, ad message exposures. *Mass Communication and Society* 11(1): 24-42.

Davies, Gary, and Takir Mian. 2010. The reputation of the party leader and of the party being led. *European Journal of Marketing* 44(3/4): 331-50.

Davis, Aeron. 2003. Public relations and news sources. In S. Cottle, ed., *News, Public Relations and Power*, 27-42. London: Sage.

Davis, Richard. 1999. *The Web of Politics: The Internet's Impact on the American Political System*. Oxford: Oxford University Press.

de Chernatony, Leslie, and Francesca Dall'Olmo Riley. 1998. Defining a "brand": Beyond the literature with experts' interpretations. *Journal of Marketing Management* 14(5): 417-43.

de Vreese, Claes H. 2005. The Spiral of Cynicism Reconsidered. *European Journal of Communication* 20(3): 283-301.

de Vreese, Claes H., and Matthijs Elenbaas. 2008. Media in the Game of Politics: Effect of Strategic Metacoverage on Political Cynicism. *International Journal of Press/Politics* 13(3): 285-309.

Deacon, David. 1996. The voluntary sector in a changing communication environment: A case study of non-official news sources. *European Journal of Communication* 11(2): 173-99.

–. 1999. Charitable images: The construction of voluntary sector news. In B. Franklin, ed., *Social Policy: The Media and Misrepresentation*, 51-69. New York: Routledge.

Dean, Jodi. 2009. Politics without politics. *Parallax* 15(3): 20-36.

Delacourt, Susan. 2012. The campaign machine. Blog. http://www.thestar.com/news/politics_blog/2012/02/the-campaign-machine.html [March 13, 2014].

Delicath, John W., and Kevin Michael DeLuca. 2003. Image events, the public sphere and argumentative practice: The case of radical environmental groups. *Argumentation* 17(3): 315-33.

Della Porta, Donatella, and Mario Diani. 1999. *Social Movements: An Introduction*. Oxford: Blackwell.

Denton Jr., Robert. 2009. Preface. In Robert E. Denton Jr., ed., *The 2008 Presidential Campaign: A Communication Perspective*, xi-xiv. Lanham, MD: Rowman and Littlefield.

Di Gennaro, Corinna, and William Dutton. 2006. The Internet and the public: Online and offline political participation in the United Kingdom. *Parliamentary Affairs* 59(2): 299-313.

Dimitrov, Roumen. 2008. The strategic response: An introduction to non-profit communication. *Third Sector Research* 14(2): 9-49.

–. 2009. New communication strategies? Lessons from the refugee rights movement in Australia. Paper presented at the 2009 Association for Non-Profit and Social Economy Research conference, Ottawa, May.

Dimmick, John. 1974. *The Gate-Keeper: An Uncertainty Theory*. Lexington, KY: Association for Education in Journalism.

Dobrzynska, Agnieszka, André Blais, and Richard Nadeau. 2003. Do the media have a direct impact on the vote? The case of the 1997 Canadian election. *International Journal of Public Opinion Research* 15(1): 27-43.

Doer, Gary. 2000. Policy challenges for the new century: The Manitoba perspective. Paper presented at the 2000 Donald Gow Lecture, Kingston, ON, April.

Doherty, Brendan J. 2007. The politics of the permanent campaign: Presidential travel and the Electoral College, 1977-2004. *Presidential Studies Quarterly* 37(4): 749-73.

Dufresne, Yannick, and Alex Marland. 2012. The Canadian political market and the rules of the game. In A. Marland, T. Giasson, and J. Lees-Marshment, eds., *Political Marketing in Canada*, 22-38. Vancouver: UBC Press.

Dulio, David A. 2004. *For Better or Worse? How Political Consultants Are Changing Elections in the United States*. Albany: State University of New York Press.

Eckstein, Harry. 1971. Case study and theory in political science. In F. Greenstein and N. Polsby, eds., *Handbook of Political Science*, vol. 7, *Strategies of Inquiry*, 119-64. Reading, MA: Addison-Wesley.

Economist. 2005. Fake news: Don't worry; It's only little brother. *Economist*, March 17. http://www.economist.com/.

Edelman, Murray J. 1985. *The Symbolic Use of Politics*. Chicago: University of Illinois Press.

Eikenberry, Angela M., and Jodie Drapal Kluver. 2004. The marketization of the non-profit sector: Civil society at risk. *Public Administration* 64(2): 132-40.

Ellul, Jacques. 1965. *Propaganda: The Formation of Men's Attitudes*. New York: Alfred A. Knopf.

Elmer, Greg, Ganaele Langlois, Zach Devereaux, Peter Ryan Malachy, Fenwick McKelvey, Joanna Redden, and Brady A. Curlew. 2009. "Blogs I read": Partisanship and party loyalty in the Canadian political blogosphere. *Journal of Information Technology and Politics* 6(2): 156-65.

Entman, Robert M. 1993. Framing: Toward clarification of a fractured paradigm. *Journal of Communication* 43(4): 51-58.

–. 2004. *Projections of Power: Framing News, Public Opinion and US Foreign Policy*. Chicago: University of Chicago Press.

–. 2007. Framing bias: Media in the distribution of power. *Journal of Communication* 57(1): 163-73.

Epstein, Edward J. 1973. *News from Nowhere: Television and the News*. New York: Random House.

Evans, Dave. 2008. *Social Media Marketing: An Hour a Day*. Indianapolis, IN: John Wiley and Sons.

Everitt, Joanna, and Michael Camp. 2009a. Changing the game changes the frame: The media's use of lesbian stereotypes in leadership versus election campaigns. *Canadian Political Science Review* 3(3): 24-39.

–. 2009b. One is not like the others: Allison Brewer's leadership of the New Brunswick NDP. In S. Bashevkin, ed., *Opening Doors Wider: Women's Political Engagement in Canada*, 127-44. Vancouver: UBC Press.

Fairclough, Norman. 1995. *Media Discourse*. London: Edward Arnold.

Falkowski, Andrzej, and Wojciech Cwalina. 2012. Political marketing: Structural models of advertising influence and voter behavior. *Journal of Political Marketing* 11(1/2): 8-26.

Fan, David P. 1988. *Predictions of Public Opinion from the Mass Media*. Westport, CT: Greenwood Press.

Farrell, David M., and Paul Webb. 2000. Political parties as campaign organizations. In Russell J. Dalton and Martin P. Wattenberg, eds., *Parties without Partisans: Political Change in Advanced Industrial Democracies*, 102-28. Oxford: Oxford University Press.

Faucheux, Ronald A. 2003. *Winning Elections: Political Campaign Management, Strategy and Tactics*. Lanham, MD: M. Evans.

Febbraro, Angela R., Michael H. Hall, and Marcus Parmegiani. 1999. *Developing a Typology of the Voluntary Health Sector in Canada: Definition and Classification Issues*. Toronto: Canadian Centre for Philanthropy.

Fiske, Susan T. 1980. Attention and weight in person perception: The impact of negative and extreme behavior. *Journal of Personality and Social Psychology* 38(6): 889-906.

Flanagan, Tom. 2007. *Harper's Team: Behind the Scenes in the Conservative Rise to Power*. Montreal and Kingston, ON: McGill-Queen's University Press.

–. 2009. *Harper's Team: Behind the Scenes in the Conservative Rise to Power*. 2nd ed. Montreal and Kingston, ON: McGill-Queen's University Press.

–. 2012. Political communication and the "permanent campaign." In D. Taras and C. Waddell, eds., *How Canadians Communicate IV: Media and Politics*, 129-48. Edmonton: Athabasca University Press.

Fleras, Augie. 2011. *The Media Gaze: Representations of Diversities in Canada*. Vancouver: UBC Press.

Fletcher, Frederick J. 1981. *The Newspaper and Public Affairs*. Volume 7 of the research studies of the Royal Commission on Newspapers. Ottawa: Minister of Supply and Services.

–. 1998. Media and political identity: Canada and Quebec in the era of globalization. *Canadian Journal of Communication* 23(3): 359-80.

Fletcher, Frederick J., and Robert Everett. 2000. The media and Canadian politics in an era of globalization. In M. Whittington and G. Williams, eds., *Canadian Politics in the 21st Century*, 5th ed., 381-402. Scarborough, ON: Nelson.

Fletcher, Frederick J., and Robert MacDermid. 1998. The rhetoric of campaign advertising in Canada: An analysis of party TV spots in 1997. Paper prepared for the annual conference of the International Association for Media and Communication Research, Glasgow, July 29.

Flood, Warren. 2013. Why big data is good for the little guy. *Hill Times*, January 21, 14.

Foster, Emilie, and Patrick Lemieux. 2012. Selling a cause: Political marketing and interest groups. In A. Marland, T. Giasson, and J. Lees-Marshment, eds., *Political Marketing in Canada*, 156-74. Vancouver: UBC Press.

Foster, Steven. 2010. *Political Communication*. Edinburgh: Edinburgh University Press.

Fourquet, Marie-Pierre, and Didier Courbet. 2004. Nouvelle méthode d'étude des cognitions en réception (ECER) et application expérimentale à la communication politique. *Revue Internationale de Psychologie Sociale* 17(3): 27-75.

Francoli, Mary, Joshua Greenberg, and Christopher Waddell. 2012. The Campaign in the Digital Media. In J.H. Pammett and C. Dornan, eds., *The Canadian Federal Election of 2011*, 219-46. Toronto: Dundurn Press.

Franz, Michael M., Paul B. Freddman, Kenneth M. Goldstein, and Travis N. Ridout. 2007. *Campaign Advertising and American Democracy*. Philadelphia: Temple University Press.

Frau-Meigs, Divina. 2001. *Mediamorphoses Américaines dans un Espace Privé Unique au Monde*. Paris: Économica.

Fridkin, Kim K., and Patrick J. Kenney. 2001. The importance of issues in Senate campaigns: Citizens' reception of issue messages. *Legislative Studies Quarterly* 26(4): 573-97.

Friends of Canadian Broadcasting. 2011. Change in parliamentary appropriation to CBC (in 2011$). http://www.friends.ca/.

Frizzell, Alan, and Anthony Westell. 1989. The media and the campaign. In A. Frizzell, J.H. Pammett, and A. Westell, eds., *The Canadian General Election of 1988*, 75-90. Ottawa: Carleton University Press.

Funk, Carolyn L. 1999. Bringing the candidate into models of candidate evaluation. *Journal of Politics* 61(3): 700-20.

Gabriel, Trip. 2012. Campaign boils down to door-to-door voter drive. *New York Times*, October 21. http://www.nytimes.com/.

Galloway, Gloria. 2006. PM fires another salvo at the press. *Globe and Mail*, April 12, A4.

Gans, Herbert J. 1979. *Deciding What's News: A Study of CBS Evening News, NBC Nightly News, Newsweek, and Time*. New York: Pantheon.

Gauthier, Gilles. 1990. Sheila Copps: Les Canadiens ont soif d'honnêteté. *La Presse*, January 16, B1.

Geer, John G. 2006. *In Defense of Negativity: Attack Ads in Presidential Campaigns*. Chicago: University of Chicago Press.

Gerbaudo, Paolo. 2012. *Tweets and the Streets: Social Media and Contemporary Activism*. London: Pluto Press.

Giasson, Thierry. 2012. As (not) seen on TV: News coverage of political marketing in Canadian federal elections. In A. Marland, T. Giasson, and J. Lees-Marshment, eds., *Political Marketing in Canada*, 175-92. Vancouver: UBC Press.

Giasson, Thierry, and David Dumouchel. 2012. Of wedge issues and conservative politics in Canada: The case of the gun registry. Paper presented at the annual meeting of the Midwest Political Science Association, Chicago, April.

Giasson, Thierry, Jennifer Lees-Marshment, and Alex Marland. 2012. Challenges for democracy. In A. Marland, T. Giasson, and J. Lees-Marshment, eds., *Political Marketing in Canada*, 241-55. Vancouver: UBC Press.

Giasson, Thierry, Vincent Raynauld, and Cyntia Darisse. 2009. The Quebec political blogosphere: A "distinct society" in North America? Paper presented at the annual meeting of the Midwest Political Science Association, Chicago, April.

–. 2011. Hypercitizens from a distinct society: Characterizing Quebec's political bloggers online and offline political involvement. *International Journal of Interactive Communication Systems and Technologies* 1(2): 29-45.

Gibson, Rachel, and Andrea Römmele. 2001. Changing campaign communications: A party-centered theory of professionalized campaigning. *Harvard International Journal of Press Politics* 6(4): 31-44.

Gidengil, Elisabeth. 2008. Media matters: Election coverage in Canada. In J. Strömbäck and L.L. Kaid, eds., *Handbook of Election Coverage around the World*, 58-72. New York: Routledge.

Gidengil, Elisabeth, André Blais, Joanna Everitt, Patrick Fournier, and Neil Nevitte. 2006. Back to the future? Making sense of the 2004 Canadian election outside Quebec. *Canadian Journal of Political Science* 39(1): 1-25.

–. 2012. *Dominance and Decline: Making Sense of Recent Canadian Elections*. Toronto: University of Toronto Press.

Gidengil, Elisabeth, André Blais, Richard Nadeau, and Neil Nevitte. 2002. Priming and campaign context: Evidence from recent Canadian elections. In D. Farrell and

R. Schmitt-Beck, eds., *Do Political Campaigns Matter? Campaign Effects in Elections and Referendums*, 71-85. London: Routledge.

–. 2003. Women to the Left? Gender differences in political beliefs and policy preferences. In M. Tremblay and L. Trimble, eds., *Gender and Electoral Representation in Canada*, 140-59. Oxford: Oxford University Press.

Gidengil, Elisabeth, André Blais, Neil Nevitte, and Richard Nadeau. 1999. Making sense of regional voting in the 1997 federal election: Liberal and Reform support outside Quebec. *Canadian Journal of Political Science* 32(2): 247-72.

–. 2004. *Citizens*. Vancouver: UBC Press.

Gidengil, Elisabeth, and Joanna Everitt. 2000. Talking tough: Gender and reported speech in campaign news coverage. Joan Shorenstein Center on the Press, Politics and Public Policy, Working Paper Series. Boston: Harvard University.

–. 2003. Conventional coverage/unconventional politicians: Gender and media coverage of Canadian leaders' debates, 1993, 1997, 2000. *Canadian Journal of Political Science* 363: 559-77.

Gieber, Walter. 1964. News is what newspapermen make it. In L.A. Dexter and D.M. White, eds., *People, Society and Mass Communication*, 172-82. New York: Free Press.

Gilens, Martin. 1999. *Why Americans Hate Welfare*. Chicago: University of Chicago Press.

Gillmor, Dan. 2006. *We the Media: Grassroots Journalism by the People, for the People*. Sebastopol, CA: O'Reilly Media.

Gingras, Anne-Marie, ed. 2003. *La Communication Politique: État des Savoirs, Enjeux et Perspectives*. Montreal: PUQ.

–. 2009. *Médias et Démocratie: Le Grand Malentendu Revue et Augmentée*. 3rd ed. Sainte-Foy, QC: Presses de L'Université du Québec.

Gingras, François-Pierre. 1995. Daily male delivery: Women and politics in the daily newspapers. In Francois-Pierre Gingras, ed., *Gender and Politics in Contemporary Canada*, 208-31. Toronto: Oxford University Press.

Gitlin, Todd. 1980. *The Whole World Is Watching: Mass Media in the Making and Unmaking of the New Left*. Berkeley: University of California Press.

Glaser, Jack, and Peter Salovey. 1998. Affect in electoral politics. *Personality and Social Psychology Review* 2(3): 156-72.

Glassman, Matthew Eric, Jacob R. Straus, and Colleen J. Shogan. 2009. Social networking and constituent communication: Member use of Twitter during a two-week period in the 111th Congress. Congressional Research Service. http://www.fas.org/.

Golbeck, Jennifer, Justin M. Grimes, and Anthony Rogers. 2010. Twitter use by the US Congress. *Journal of the American Society for Information Science and Technology* 61(8): 1612-21.

Goldenberg, Eddie. 2006. *The Way It Works: Inside Ottawa*. Toronto: McClelland and Stewart.

Goldstein, Ken, and Paul Freedman. 2002. Campaign advertising and voter turnout: New evidence for a stimulation effect. *Journal of Politics* 64(3): 721-40.

Goldstein, Kenneth, and Travis N. Ridout. 2004. Measuring the effects of televised political advertising in the United States. *Annual Review of Political Science* 7(1): 205-26.

Goodyear-Grant, Elizabeth. 2009. Crafting a public image: Women MPs and the dynamics of media coverage. In S. Bashevkin, ed., *Opening Doors Wider: Women's Political Engagement in Canada,* 147-66. Vancouver: UBC Press.

Gormley Jr., William T., and Helen Cymrot. 2006. The strategic choices of child advocacy groups. *Non-Profit and Voluntary Sector Quarterly* 35(1): 102-22.

Gossage, Patrick. 1986. *Close to the Charisma: My Years between the Press and Pierre Elliot Trudeau.* Toronto: McClelland and Stewart.

Gould, Phillip. 2002. What permanent campaign? BBC News, June 4. http://news.bbc.co.uk/.

Government of Canada. 2001. *Telephone Directory, Ottawa-Hull.* Ottawa: Department of Communications.

–. 2006. *Telephone Directory, Ottawa-Hull.* Ottawa: Public Works and Government Services Canada.

Grabe, Maria Elizabeth, and Eric Page Bucy. 2009. *Image Bite Politics: News and the Visual Framing of Elections.* Oxford: Oxford University Press.

Graber, Doris, Denis McQuail, and Pippa Norris. 1998. Introduction: Political communication in a democracy. In Doris Graber, Denis McQuail, and Pippa Norris, eds., *The Politics of News, The News of Politics,* 1-16. Washington, DC: CQ Press.

Grant, Will, Brenda Moon, and Janie Busby Grant. 2010. Digital dialogue? Australian politicians' use of the social network tool Twitter. *Australian Journal of Political Science* 45(4): 579-604.

Green-Pedersen, Christoffer, and Kees van Kersbergen. 2002. The politics of the "third way": The transformation of social democracy in Denmark and the Netherlands. *Party Politics* 8(5): 507-24.

Greenberg, Joshua, and Georgina Grosenick. 2008. Communicative capacity in the third sector. *Third Sector Review* 14(2): 51-74.

Greenberg, Joshua, and Maggie MacAulay. 2009. NPO 2.0: Exploring the web presence of environmental nonprofit organizations in Canada. *Global Media Journal* 2(1): 63-88.

Greenberg, Joshua, and David Walters. 2004. Promoting philanthropy? News publicity and voluntary organizations in Canada. *Voluntas* 15(4): 383-404.

Greenwald, Anthony G. 1968. Cognitive learning, cognitive response to persuasion, and attitude change. In A.G. Greenwald, T.C. Brock, and T.M. Ostrom, eds., *Psychological Foundations of Attitudes,* 147-70. San Diego: Academic Press.

Grenier, Eric. 2013. Why were the polls completely wrong about the BC election? *Globe and Mail,* May 15. http://www.theglobeandmail.com/.

Gurevitch, Michael, and Jay G. Blumler. 1990. Political communication systems and democratic values. In J. Lichtenberg, ed., *Democracy and the Mass Media,* 24-35. Cambridge: Cambridge University Press.

Guthey, Eric, and Brad Jackson. 2005. CEO portraits and the authenticity paradox. *Journal of Management Studies* 42(5): 1057-82.

Gutmann, Amy, and Dennis Thompson. 1996. *Democracy and Disagreement.* Cambridge, MA: Harvard University Press.

Hackett, Robert A. 1991. *News and Dissent: The Press and Politics of Peace in Canada.* Norwood, NJ: Ablex.

–. 2001. News media's influence on Canadian party politics: Perspectives on a shifting relationship. In H.G. Thorburn and A. Whitehorn, eds., *Party Politics in Canada*, 8th ed., 381-97. Toronto: Prentice Hall.

–. 2005. Is there a democratic deficit in US and UK journalism? In Stuart Allan, ed., *Journalism: Critical Issues*, 85-97. New York: Open University Press.

Hallin, Daniel C., and Paolo Mancini. 2004. *Comparing Media Systems: Three Models of Media and Politics*. Cambridge: Cambridge University Press.

Harell, Allison, Stuart N. Soroka, Shanto Iyengar, and Nicholas A. Valentino. 2012. The impact of economic and cultural cues on support for immigration in Canada and the US. *Canadian Journal of Political Science* 45(3): 499-530.

Harell, Allison, Stuart Soroka, and Kiera Ladner. 2012. Public opinion, prejudice and the racialization of welfare in Canada. Paper presented at the annual meeting of the Canadian Political Science Association, Edmonton, June.

Harell, Allison, Stuart N. Soroka, and Adam Mahon. 2008. Is welfare a dirty word? Canadian public opinion on social assistance policies. *Policy Options* 29(8): 53.

Harris Decima. 2010. Majority want Conservatives and Liberals to get new leaders. *Harris Decima*, November 17. http://www.harrisdecima.com/.

Harrison, Riva. 2011. Former communications director, Manitoba NDP. Personal communication, August 31.

Hart, Paul 't, and Karen Tindall. 2009. Leadership by the famous: Celebrity as political capital. In J. Kane, H. Patapan, and P. 't Hart, eds., *Dispersed Democratic Leadership: Origins, Dynamics and Implications*, 255-78. Oxford: Oxford University Press.

Hart, Ted, James M. Greenfield, and Michael Johnston. 2005. *Non-Profit Internet Strategies: Best Practices for Marketing, Communication and Fundraising Success*. Hoboken, NJ: John Wiley and Sons.

Heath, Anthony F., Roger M. Jowell, and John K. Curtice. 2001. *The Rise of New Labour: Party Policies and Voter Choices*. Oxford: Oxford University Press.

Heck, Ronald H. 2004. *Studying Educational and Social Policy: Theoretical Concepts and Research Methods*. Mahwah, NJ: Erlbaum.

Heclo, Hugh. 2000. Campaigning and governing: A conspectus. In N. Ornstein and T. Mann, eds., *The Permanent Campaign and Its Future*, 1-37. Washington, DC: American Enterprise Institute and Brookings Institution.

Heldman, Caroline, Susan J. Carroll, and Stephanie Olson. 2005. "She brought only a skirt": Print media coverage of Elizabeth Dole's bid for the Republican presidential nomination. *Political Communication* 22(3): 15-35.

Hellström, Åke, and Joseph Tekle. 1994. Person perception through facial photographs: Effects of glasses, hair, and beard on judgments of occupation and personal qualities. *European Journal of Social Psychology* 24(6): 693-705.

Hendricks John Allen, and Jerry K. Frye. 2011. Social media and the millennial generation in the 2010 Midterm Election. In H.S. Noor Al-Deen and J.A. Hendricks, eds., *Social Media: Usage and Impact*, 183-200. Lanham, MD: Lexington Books.

Herman, Edward S., and Noam Chomsky. 1988. *Manufacturing Consent: The Political Economy of Mass Media*. New York: Pantheon.

Hertog, James K., and Douglas M. McLeod. 2001. A multiperspectival approach to framing analysis: A field guide. In S.D. Reese Jr., O.H. Gandy, and A.E. Grant, eds.,

Framing Public Life: Perspectives on Media and Our Understanding of the Social World, 141-62. London: Lawrence Erlbaum Associates.

Hess, Steven. 2000. The press and the permanent campaign. In N. Ornstein and T. Mann, eds., *The Permanent Campaign and Its Future*, 38-53. Washington, DC: American Enterprise Institute and Brookings Institution.

Holtz-Bacha, Christina. 2002. Professionalization of political communication. *Journal of Political Marketing* 1(4): 23-37.

Honeycutt, Courtenay, and Susan C. Herring. 2009. Beyond microblogging: Conversation and collaboration via Twitter. *Proceedings of the Forty-Second Hawai'i International Conference on System Sciences*. Los Alamitos, CA: IEEE Press.

Howard, Philip N. 2006. *New Media Campaigns and the Managed Citizen*. Cambridge: Cambridge University Press.

Huffington Post. 2014. Stephen Harper's "24 Seven" show has North Korean vibes. Video. January 9. http://www.huffingtonpost.ca.

Huffington Post BC. 2013. BC election 2013 polls wrong, pollsters lambasted. May 15. http://www.huffingtonpost.ca/.

Ibbitson, John. 2009. Lacrosse trumps torture for Stephen Harper. Ottawa notebook weblog. *Globe and Mail*, November 23. http://www.theglobeandmail.com/.

–. 2012. Like Energizer bunny, PM could go on and on. *Globe and Mail*, July 23. http://www.theglobeandmail.com/.

Imagine Canada. 2006. National survey of non-profit and voluntary organizations: The non-profit and voluntary sector in Canada. Toronto.

–. 2012. Charities engagement in public policy. http://us1.campaign-archive1.com/.

Interbrand. 2010. Best Canadian brands 2010. http://www.interbrand.com/.

Ipsos. 2011. Canada's love affair with online social networking continues. July 14. http://www.ipsos-na.com/news-polls/pressrelease.aspx?id=5286.

Iyengar, Shanto. 1990. Framing responsibility for political issues: The case of poverty. *Political Behaviour* 12(1): 19-40.

–. 1991. *Is Anyone Responsible? How Television Frames Political Issues*. Chicago: University of Chicago Press.

–. 1997. Overview. In S. Iyengar and R. Reeves, eds., *Do the Media Govern? Reporters, Politicians and the American People*, 211-16. London: Sage.

Iyengar, Shanto, and Donald R. Kinder. 1987. *News That Matters: Television and American Opinion*. Chicago: University of Chicago Press.

Jacobs, Ronald N., and Daniel J. Glass. 2002. Media publicity and the voluntary sector: The case of non-profit organizations in New York City. *Voluntas* 13(3): 235-52.

Jacobson, Gary C. 2007. *A Divider, Not a Uniter: George W. Bush and the American People*. New York: Pearson Longman.

Jamieson, Doug. 2009. Building relationships in the networked age: Some implications of the Internet for nonprofit organizations. *Philanthropist* 15(2): 23-32.

Jamieson, Kathleen Hall. 1995. *The Double Bind: Women and Leadership*. New York: Oxford University Press.

Jankowski, Nicholas W., and Martine Van Selm. 2008. Internet-based political communication research: Illustrations, challenges and innovations. *The Public* 15(2): 5-16.

Jansen, Harold J., and Royce Koop. 2005. Pundits, ideologues, and the ranters: The British Columbia election online. *Canadian Journal of Communication* 30(4): 613-32.

Jin, Hyun S., Soontae An, and Todd Simon. 2009. Beliefs of and attitudes toward political advertising: An exploratory investigation. *Psychology and Marketing* 26(6): 551-68.

Johansen, Helen P.M. 2012. *Relational Political Marketing in Party-Centred Democracies.* Farnham, UK: Ashgate.

Johnson, Dennis W. 2011. *Campaigning in the Twenty-First Century: A Whole New Ballgame?* New York: Routledge.

Johnson, Ron. 2011. Co-founder and former president of Now Communications. Personal communication, August 24.

Johnson, Thomas J., and Barbara K. Kaye. 2004. Wag the blog: How reliance on traditional media and the Internet influence credibility perceptions of weblogs among blog users. *Journalism and Mass Communication Quarterly* 81(3): 622-42.

Johnson, William. 2005. *Stephen Harper and the Future of Canada.* Toronto: McClelland and Stewart.

Johnston, Richard, André Blais, Henry E. Brady, and Jean Crête. 1992. *Letting the People Decide: Dynamics of a Canadian Election.* Montreal and Kingston, ON: McGill-Queen's University Press.

Johnston, Richard, and Henry E. Brady. 2002. The rolling cross-section design. *Electoral Studies* 21(2): 283-95.

Johnston, Richard, Michael G. Hagen, and Kathleen H. Jamieson. 2004. *The 2000 Presidential Election and the Foundations of Party Politics.* Cambridge: Cambridge University Press.

Juslin, Patrick N. 2001. Communicating emotion in music performance: A review and a theoretical framework. In P.N. Juslin and J.A. Sloboda, eds., *Music and Emotions*, 309-37. New York: Oxford University Press.

Kaase, Max. 1994. Is there personalization in politics? Candidates and voting behavior in Germany. *International Political Science Review* 15(3): 211-30.

Kaid, Lynda Lee. 2000. Ethics in political advertising. In R.E. Denton Jr., ed., *Political Communication Ethics,* 146-77. Westport, CT: Praeger.

–. 2012. Political advertising as political marketing: A retro-forward perspective. *Journal of Political Marketing* 11(1-2): 29-53.

Kaid, Lynda L., Monica Postelnicu, Kristen Landreville, Hyun J. Yun, and Abby G. LeGrange. 2007. The effects of political advertising on young voters. *American Behavioral Scientist* 50(9): 1137-51.

Kam, Chris. 2001. Do ideological preferences explain parliamentary behaviour? Evidence from Great Britain and Canada. *Journal of Legislative Studies* 7(4): 89-126.

Kam, Cindy. 2005. Who toes the party line? Cues, values, and individual differences. *Political Behavior* 27(2): 163-82.

Karvonen, Lauri. 2010. *The Personalisation of Politics: A Study of Parliamentary Democracies.* Essex, UK: ECPR Press.

Katz, Richard S. 1997. *Democracy and Elections.* Oxford: Oxford University Press.

Kaufmann, Karen M. 2004. Disaggregating and reexamining issue ownership and voter choice. *Polity* 36(2): 283-99.

Kavanagh, Dennis. 1995. *Election Campaigning: The New Marketing of Politics*. Oxford: Blackwell.

Keeter, Scott. 1987. The illusion of intimacy television and the role of candidate personal qualities in voter choice. *Public Opinion Quarterly* 51(3): 344-58.

Kenix, Linda Jean. 2008. Nonprofit organizations' perceptions and uses of the Internet. *Television and New Media* 9(5): 407-28.

Kennedy, Mark. 2010. Tories strong on jobs, crime; soft on health care, pensions: poll. *National Post*, December 19. http://www.nationalpost.com/news/.

Kensicki, Linda Jean. 2004. No cure for what ails us: The media-constructed disconnect between societal problems and possible solutions. *Journalism and Mass Communication Quarterly* 81(1): 53-73.

Kernaghan, Kenneth. 2007. Moving beyond politics as usual? Online campaigning. In S. Borins, K. Kernaghan, D. Brown, N. Bontis, and F. Thompson, eds., *Digital State at the Leading Edge*, 183-223. Toronto: University of Toronto Press.

Kernell, Samuel. 1997. *Going Public: New Strategies of Presidential Leadership*. 3rd ed. Washington, DC: Congressional Quarterly.

Key, V.O. 1966. *The Responsible Electorate: Rationality in Presidential Voting, 1936-1960*. Cambridge: Belknap Press of Harvard University Press.

Kietzmann, Jan, Kristopher H. Hermkens, Ian P. McCarthy, and Bruno S. Silvestre. 2011. Social media? Get serious! Understanding the functional building blocks of social media. *Business Horizons* 54(3): 241-51.

Kingdon, John. 1984. *Agenda-Setting, Alternatives and Public Policies*. New York: HarperCollins.

Kitschelt, Herbert. 1994. *The Transformation of European Social Democracy*. Cambridge: Cambridge University Press.

Klein, Naomi. 2000. *No Logo: Taking Aim at the Brand Bullies*. Toronto: Vintage Canada.

Klodawsky, Fran. 2004. Tolerating homelessness in Canada's capital: Gender, place and human rights. *HAGAR: Studies in Culture, Polity and Identities* 5(1): 103-18.

Knobloch-Westerwick, Nikhil, Silvia Sharma, Derek L. Hansen, and Scott Alter. 2005. Impact of popularity indications on readers' selective exposure to online news. *Journal of Broadcasting and Electronic Media* 49(3): 296-313.

Koop, Royce. 2012. Marketing and efficacy: Does political marketing empower Canadians? In A. Marland, T. Giasson, and J. Lees-Marshment, eds., *Political Marketing in Canada*, 224-38. Vancouver: UBC Press.

Koop, Royce, and Harold J. Jansen. 2006. Canadian political blogs: Online soapboxes or forums for democratic dialogue? Paper presented at the annual meeting of the Canadian Political Science Association, Toronto, May.

–. 2009. Political blogs and blogrolls in Canada: Forums for democratic deliberation? *Social Science Computer Review* 27(2): 155-73.

Kostyra, Eugene. 2011. Former cabinet minister and elections target riding director, Manitoba NDP. Personal communication, September 2.

Kozolanka, Kirsten. 2006. The sponsorship scandal as communication: The rise of politicized and strategic communication in the federal government. *Canadian Journal of Communication* 31(2): 343-66.

–. 2009. Communication by stealth: The new common sense in government communication. In A.M. Maslove, ed., *How Ottawa Spends, 2009-2010: Economic Upheaval and Political Dysfunction*, 222-40. Montreal and Kingston, ON: McGill-Queen's University Press.

–. 2012. "Buyer" beware: Pushing the boundaries of marketing communications in government. In A. Marland, T. Giasson, and J. Lees-Marshment, eds., *Political Marketing in Canada*, 107-22. Vancouver: UBC Press.

Kozolanka, Kirsten, Patricia Mazepa, and David Skinner, eds. 2012. *Alternative Media in Canada*. Vancouver: UBC Press.

Kramer, Ralph M. 1981. *Voluntary Agencies in a Welfare State*. Berkeley: University of California Press.

Kriesi, Hanspeter. 2012. Personalization of national election campaigns. *Party Politics* 18(6): 825-44.

Krueger, Brian S. 2002. Assessing the potential of Internet political participation in the United States: A resource approach. *American Politics Research* 30(5): 476-98.

Kullin, Hans. 2008. BlogSweden 3: A survey of Swedish bloggers and blog readers. http://www.kullin.net/.

Lalancette, Mireille, and Catherine Lemarier-Saulnier. 2013. Gender and political evaluation in leadership races. In R. Lexier and T. Small, eds., *Mind the Gaps: Canadian Perspectives on Gender and Politics*, 116-30. Halifax: Fernwood Press.

Lang, Kurt, and Gladys Engel Lang. 1966. The mass media and voting. In B. Berelson and M. Janowitz, eds., *Reader in Public Opinion and Communication*, 455-72. New York: Free Press.

Langer, Anà-Inès. 2010. The politicization of private persona: Exceptional leaders or the new rule? The case of the United Kingdom and the Blair effect. *International Journal of Press/Politics* 15(1): 60-76.

–. 2012. *The Personalisation of Politics in the UK: Mediated Leadership from Attlee to Cameron*. Manchester: Manchester University Press.

Laporte, Stéphane. 2003. La dalaï-Sheila. *La Presse*, September 28, A5.

Lau, Richard R., and David P. Redlawsk. 2006. *How Voters Decide: Information Processing during Election Campaigns*. Cambridge: Cambridge University Press.

Lau, Richard R., Lee Sigelman, Caroline Heldman, and Paul Babbitt. 1999. The effects of negative political advertisements: A meta-analytic assessment. *American Political Science Review* 93(4): 851-75.

Lau, Richard R., Lee Sigelman, and Ivy Brown Rovner. 2007. The effect of negative political campaigns: A meta-analytic reassessment. *Journal of Politics* 69(4): 1176-1209.

Lavigne, Brad. 2012. Former national director and campaign director, federal NDP. Personal communication, March 8.

–. 2013. Personal correspondence, January 18.

Lawless, Jennifer L. 2012. Twitter and Facebook: New ways to send the same old message? In R.L. Fox and J. Ramos, eds., *iPolitics: Citizens, Elections, and Governing in the New Media Era*, 206-32. New York: Cambridge University Press.

Lawrence, Regina, and Melody L. Rose. 2010. *Hillary Clinton's Race for the White House: Gender Politics and the Media on the Campaign Trail*. Boulder, CO: Lynne Rienner.

Lazarsfeld, Paul F., Bernard Berelson, and Hazel Gaudet. 1948. *The People's Choice: How the Voter Makes up His Mind in a Presidential Campaign.* New York: Columbia University Press.

Leclerc, Bernard-Simon, and Clément Dassa. 2010. Interrater reliability in content analysis of healthcare service quality using Montreal's conceptual framework. *Canadian Journal of Program Evaluation* 24(2): 81-102.

Lee, Eun-Ju, and Yoon Jae Jang. 2010. What do others' reactions to news on Internet portal sites tell us? Effects of presentation format and readers' need for cognition on reality perceptions. *Communication Research* 37(6): 825-46.

Lee, Jae Kook. 2007. The effect of the Internet on homogeneity of the media agenda: A test of the fragmentation thesis. *Journalism and Mass Communication Quarterly* 84(4): 745-60.

Lees-Marshment, Jennifer. 2001. The product, sales and market-oriented party: How Labour learnt to market the product, not just the presentation. *European Journal of Marketing* 35(9): 1074-84.

Lemarier-Saulnier, Catherine, and Mireille Lalancette. 2012. La dame de fer, la bonne mère et les autres: Une analyse du cadrage médiatique de politiciennes Canadiennes. *Canadian Journal of Communication* 37(3): 459-86.

Levine, Allan. 1993. *Scrum Wars: The Prime Ministers and the Media.* Toronto: Dundurn Press.

Lewin, Kurt. 1947. Frontiers in group dynamics II: Channels of group life; social planning and action research. *Human Relations* 1(2): 143-53.

Liberal Democrats. 2013. Connect. http://www.libdems.org.uk/connect.aspx.

Liberal Party of Canada. 2011. "Building a modern Liberal party: A background paper for discussion among members of the Liberal party." http:convention.liberal.ca/files/2011/12/BuildingModernLiberalPartyFinal.pdf

–. 2013a. About Liberalist. https://liberalist.liberal.ca/.

–. 2013b. Account types. http://liberalist.liberal.ca/.

–. 2013c. Available courses: Liberalist champions. http://classe.liberaliste.ca/.

–. 2013d. What Is Liberalist? http://liberalist.liberal.ca/.

Libin, Kevin. 2006. "I've got more control now." Exclusive interview: Prime Minister Stephen Harper explains why the press gallery's fighting him – and why it's backfiring. *Western Standard,* June 19. http://www.westernstandard.ca/.

Lilleker, Darren G. 2006. *Key Concepts in Political Communication.* London: Sage.

Lloyd, Jenny. 2006. The 2005 general election and the emergence of the "negative brand." In D.G. Lilleker, N.A. Jackson, and R. Scullion, eds., *The Marketing of Political Parties: Political Marketing at the 2005 British General Election,* 59-80. Manchester: Manchester University Press.

–. 2009. After Blair: The challenge of communicating Brown's brand of Labour. In J. Lees-Marshment, ed., *Political Marketing: Principles and Applications,* 232-35. New York: Routledge.

Loader, Brian D., and Dan Mercea. 2011. Networking democracy? Social media innovations and participatory politics. *Information, Communication and Society* 14(6): 757-69.

Longford, Graham. 2002. Canadian democracy hard-wired? Connecting government and citizens in the digital age. *Canadian Issues,* June: 33-38.

Lovejoy, Kristen, Richard Waters, and Gregory D. Saxton. 2012. Engaging stakeholders through Twitter: How nonprofit organizations are getting more out of 140 characters or less. *Public Relations Review* 38(2): 313-18.

Low, George S., and Charles W. Lamb Jr. 2000. The measurement and dimensionality of brand associations. *Journal of Product and Brand Management* 9(6): 350-70.

MacDermott, Kathy. 2008. Marketing government: The public service and the permanent campaign. Report No. 10 (1-118). Melbourne: School of Social Sciences, Australian National University.

Macdonald, Myra. 1998. Personalization in current affairs journalism. *The Public* 5(3): 109-26.

MacGregor, Robert M. 2003. I am Canadian: National identity in beer commercials. *Journal of Popular Culture* 37(2): 276-86.

MacKuen, Michael B., and Steven L. Coombs. 1981. *More than News: Media Power in Public Affairs.* Beverly Hills: Sage.

Maclean's. 2012. 2012 advertising rates. http://www.rogersconnect.com/.

MacSween, Mike. 2011. Former provincial secretary, Nova Scotia NDP. Personal communication, September 1.

Mahon, Adam. 2009. What lies beyond the gates: Media gatekeepers and the framing of welfare news in Canada. MA thesis, McGill University.

Manning, Paul. 2001. *News and News Sources: A Critical Introduction.* Thousand Oaks, CA: Sage.

Marcus, Georges E., Russell W. Neuman, and Michael B. MacKuen. 2000. *Affective Intelligence and Political Judgment.* Chicago: University of Chicago Press.

Marland, Alex. 2012. Political photography, journalism and framing in the digital age: The management of visual media by the prime minister of Canada. *International Journal of Press/Politics* 17(2): 214-33.

Marland, Alex, and Tom Flanagan. 2013. Brand new party: Political branding and the Conservative Party of Canada. *Canadian Journal of Political Science* 46(4): 951-72.

Marland, Alex, Thierry Giasson, and J. Lees-Marshment, eds. 2012. *Political Marketing in Canada.* Vancouver: UBC Press.

Marshall, David P. 1997. *Celebrity and Power.* Minneapolis: University of Minnesota Press.

Martin, Lawrence. 2003. *Iron Man: The Defiant Reign of Jean Chrétien.* Toronto: Viking Canada.

–. 2006. A conservative Pierre Trudeau is taking charge. *Globe and Mail,* February 23, A19.

–. 2010. *Harperland: The Politics of Control.* Toronto: Viking Canada.

Martin, Shauna. 2011. Former researcher/director/health minister's chief of staff, Manitoba NDP and director of strategic operations, Nova Scotia NDP. Personal communication, September 29.

Marwick, Alice E., and Danah Boyd 2011. I tweet honestly, I tweet passionately: Twitter users, context collapse, and the imagined audience. *New Media and Society* 13(1): 114-33.

Mayer, Jeremy D. 2004. The presidency and image management: Discipline in pursuit of illusion. *Presidential Studies Quarterly* 34(3): 620-31.

Mayfield, Anthony. 2008. *What Is Social Media?* iCrossing (ebook). http://www.icrossing. co.uk/.

Mazzoleni, Gianpietro, and Winfried Schulz. 1999. Mediatization of politics: A challenge for democracy? *Political Communication* 16(3): 247-61.

McAllister, Ian. 2007. The personalization of politics. In Russell J. Dalton and Hans-Dieter Klingemann, eds., *The Oxford Handbook of Political Behaviour*, 571-88. Oxford: Oxford University Press.

McChesney, Robert W. 1999. *Rich Media, Poor Democracy: Communication Politics in Dubious Times*. Chicago: University of Illinois Press.

McCombs, Maxwell E. 1997. Building consensus: The news media's agenda-setting roles. *Political Communication* 14(4): 433-43.

–. 2004. *Setting the Agenda: The Mass Media and Public Opinion*. Cambridge, UK: Polity Press.

–. 2005. A look at agenda-setting: Past, present and future. *Journalism Studies* 6(4): 543-57.

McCombs, Maxwell, and Donald L. Shaw. 1972. The agenda-setting function of mass media. *Public Opinion Quarterly* 36(2): 176-87.

McGrane, David. 2011. Political marketing and the NDP's historical breakthrough. In J. Pammett and C. Dornan, eds., *The Canadian Federal Election of 2011*, 77-110. Toronto: Dundurn Press.

McGrath, Anne. 2012. Former chief of staff, federal NDP. Personal communication, March 1.

McGregor, Janyce. 2012. How parties "identify" voters, and why it matters. CBC News, March 2. http://www.cbc.ca/news/.

McGuire, William J. 1964. Inducing resistance to persuasion: Some contemporary approaches. In L. Berkowitz, ed., *Advances in Experimental Social Psychology*, vol. 1, 191-229. New York: Academic Press.

–. 1989. Theoretical foundations of campaigns. In R.E. Rice and C.K. Atkin, eds., *Public Communication Campaigns*, 2nd ed., 43-65. Newbury Park, CA: Sage.

McKenna, Laura, and Antoinette Pole. 2004. Do blogs matter? Weblogs in American politics. Paper presented at the annual meeting of the American Political Science Association, Chicago, September.

–. 2007. What do bloggers do: An average day on an average political blog. *Public Choice* 134(1): 97-108.

McLaughlin, Audrey. 1992. *A Woman's Place: My Life and Politics*. Toronto: Macfarlane Walter and Ross.

McNair, Brian. 2007. Pressure group politics and the oxygen of publicity. In B. McNair, ed., *An Introduction to Political Communication*, 163-86. London: Routledge.

–. 2009. The Internet and changing global media environment. In A. Chadwick and P. Howard, eds., *Routledge Handbook of Internet Politics*, 217-29. London: Routledge.

–. 2011. *An Introduction to Political Communication*. London: Routledge.

McNelly, John T. 1959. Intermediary communicators in the international flow of news. *Journalism Quarterly* 36(1): 23-6.

Meirick, Patrick. 2002. Cognitive response to negative and comparative political advertising. *Journal of Advertising* 31(1): 49-62.

Mendelsohn, Matthew. 1993. Television's frames in the 1988 Canadian election. *Canadian Journal of Communication* 18(2): 149-71.

–. 1994. The media's persuasive effects: The priming of leadership in the 1988 Canadian election. *Canadian Journal of Political Science* 27(1): 81-97.

–. 1996. The media and interpersonal communications: The priming of issues, leaders, and party identification. *Journal of Politics* 58(1): 112-25.

Miljan, Lydia, and Barry Cooper. 2003. *Hidden Agendas: How Journalists Influence the News.* Vancouver: UBC Press.

Miller, M. Mark. 1997. Frame mapping and analysis of news coverage of contentious issues. *Social Science Computer Review* 15(4): 367-78.

Miller, M. Mark, and Bonnie P. Riechert. 1994. Identifying themes via concept mapping: A new method of content analysis. Presented at the annual meeting of the Communication Theory and Methodology Division of the Association for Education in Journalism and Mass Communication, Atlanta, August.

Mirchandani, Kiran, and Wendy Chan. 2007. *Criminalizing Race, Criminalizing Poverty: Welfare Fraud Enforcement in Canada.* Winnipeg: Fernwood Press.

Morris, Jonathan S., and Rosalee A. Clawson. 2005. Media coverage of Congress in the 1990s: Scandals, personalities, and the prevalence of policy and process. *Political Communication* 22(3): 297-313.

Mossberger, Karen, Caroline J. Tolbert, and Ramona McNeal. 2008. *Digital Citizenship: The Internet, Society, and Participation.* Cambridge, MA: MIT Press.

Murray, Catherine. 2007. The media. In L. Dobuzinskis, M. Howlett, and D. Laycock, eds., *Policy Analysis in Canada: The State of the Art,* 525-50. Toronto: University of Toronto Press.

Mutz, Diana C. 1998. *Impersonal Influence: How Perceptions of Mass Collectives Affect Political Attitudes.* New York: Cambridge University Press.

Nadeau, Richard, and Thierry Giasson. 2005. Canada's democratic malaise: Are the media to blame? In P. Howe, R. Johnston, and A. Blais, eds., *Strengthening Canadian Democracy,* 229-68. Montreal: IRPP.

Nanos, Nik. 2007. SES research poll – best PM. http://www.nikonthenumbers.com/.

Naumetz, Tim. 2011. Tory ad featuring Harper in partisan manner. *Hill Times,* January 19. http://www.hilltimes.com/.

Neatby, H. Blair. 1973. *Laurier and a Liberal Quebec: A Study in Political Management.* Toronto: McClelland and Stewart.

Needham, Catherine. 2005. Brand leaders: Clinton, Blair and the limitations of the permanent campaign. *Political Studies* 53(2): 343-61.

Negrine, Ralph, and Darren Lilleker. 2002. The professionalization of political communication: Continuities and changes in media practices. *European Journal of Communication* 17(3): 305-23.

Nelson, Barbara. 1984. *Making an Issue of Child Abuse: Political Agenda-Setting for Social Problems.* Chicago: University of Chicago Press.

Nesbitt-Larking, Paul. 2001. *Politics, Society and the Media: Canadian Perspectives.* Peterborough, ON: Broadview Press.

–. 2009. Reframing campaigning: Communications, the media and elections in Canada. *Canadian Political Science Review* 3(2): 5-22.

Nesbitt-Larking, Paul W., and Jonathan Rose. 2004. Political advertising in Canada. In D.A. Schultz, ed., *Lights, Camera, Campaign! Media, Politics, and Political Advertising*, 273-99. New York: Peter Lang.

Neustadt, Richard E. 1960. *Presidential Power: The Politics of Leadership*. New York: John Wiley and Sons.

Neveu, Érik. 2005. Politicians without politics, a polity without citizens: The politics of the chat show in contemporary France. *Modern and Contemporary France* 13(3): 323-35.

Nevitte, Neil, André Blais, Elisabeth Gidengil, and Richard Nadeau. 2000. *Unsteady State: The 1997 Canadian Federal Election*. Don Mills, ON: Oxford University Press.

Newman, Bruce I. 1999. *The Mass Marketing of Politics: Democracy in an Age of Manufactured Images*. Thousand Oaks, CA: Sage.

Newton, Kenneth. 1999. Mass media effect: Mobilization or media malaise? *British Journal of Political Science* 29(4): 577-99.

Nicholson, Stephen. 2011. Dominating cues and the limits of elite influence. *Journal of Politics* 73(4): 1165-77.

Nielson, Rasmus Keis. 2012. *Ground Wars: Personalized Communication in Political Campaigns*. Princeton, NJ: Princeton University Press.

Nip, Joyce Y. 2006. Exploring the second phase of public journalism. *Journalism Studies* 7(2): 212-36.

Noelle-Neumann, Elisabeth. 1974. The spiral of silence: A theory of public opinion. *Journal of Communication* 24(2): 43-51.

Nolan, Michael. 1981. Political communication methods in Canadian federal election campaigns. *Canadian Journal of Communication* 7(4): 28-46.

Norris, Pippa. 1997a. Women leaders worldwide: A splash of color in the photo op. In P. Norris, ed., *Women, Media and Politics*, 148-65. Oxford: Oxford University Press.

–. 1997b. *Women, Media and Politics*. Oxford: Oxford University Press.

–. 2000. *A Virtuous Circle: Political Communication in Postindustrial Societies*. Cambridge: Cambridge University Press.

–. 2001. *Digital Divide: Civic Engagement, Information Poverty, and the Internet Worldwide*. Cambridge: Cambridge University Press.

Norris, Pippa, John Curtice, David Sanders, Margaret Scammell, and Holli A. Semetko. 1999. *On Message: Communicating the Campaign*. London: Sage.

Nova Scotia NDP. 1998. Priorities for people. http://www.poltext.capp.ulaval.ca/.

–. 2006. Better deal 2006: The plan you can count on. http://www.poltext.capp.ulaval.ca/.

–. 2009. Better deal 2009: The NDP plan to make life better for today's families. http://www.poltext.capp.ulaval.ca/.

O'Connor, Dan. 2011. Premier's chief of staff, Nova Scotia NDP. Personal communication, September 19.

O'Neil, Brenda, and David Stewart. 2009. Gender and political party leadership in Canada. *Party Politics* 15(6): 737-57.

O'Shaughnessy, Nicholas J., and Stephan C. Henneberg, eds. 2002. *The Idea of Political Marketing*. London: Praeger.

Office of the Auditor General of Ontario. 2012a. *2012 Annual Report*. http://www. auditor.on.ca/.

–. 2012b. What we do. http://www.auditor.on.ca/.

Office of the Prime Minister. 2013. Photo of the day. http://www.pm.gc.ca.

Ontario. 2004. Government Advertising Act, 2004. http://www.e-laws.gov.on.ca/.

Ornstein, Michael D., H. Michael Stevenson, and A. Paul Williams. 1980. Region, class and political culture in Canada. *Canadian Journal of Political Science* 13(2): 227-71.

Ottawa Citizen. 2008. Checking the mail: MPs to get refresher on postage privilege. February 4. http://www.canada.com/ottawacitizen/.

Page, Ruth E. 2003. "Cherie: lawyer, wife, mum": Contradictory patterns of representation in media reports of Cherie Booth/Blair. *Discourse and Society* 14(5): 559-79.

Papacharissi, Ziri A. 2010. *A Private Sphere: Democracy in a Digital Age*. Cambridge, UK: Polity.

Paquin, Gilles. 1989. Le Lac Meech embête le congrès du NPD. *La Presse*, December 1, A1.

Parker, George. 2012. Chairman of the British press gallery, 2010. Personal communication, April 3.

Parliament of Canada. 2011. Member's expenditures report, April 1, 2010, to March 31, 2011. http://www.parl.gc.ca/.

Parmelee, John, and Shannon Bichard. 2012. *Politics and the Twitter Revolution: How Tweets Influence the Relationship Between Political Leaders and the Public*. Lanham, MD: Lexington Books.

Parry-Giles, Shawn. 2001. Political authenticity, television news, and Hillary Rodham Clinton. In R.P. Hart and B.H. Sparrow, eds., *Politics, Discourse, and American Society: New Agendas*, 193-210. New York: Rowman and Littlefield.

Pease, Andrew, and Paul Brewer. 2008. The Oprah factor: The effects of a celebrity endorsement in a presidential primary campaign. *International Journal of Press/Politics* 13(4): 386-400.

Pedersen, Sarah, and Caroline Macafee. 2007. Gender differences in British blogging. *Journal of Computer-Mediated Communication* 12(4): 1472-92.

Pelletier, Francine. 1990. Passe la téquila, Sheila. *La Presse*, March 10, B3.

Penn, Gemma. 2000. Semiotic analysis of still images. In M.W. Bauer and G. Gaskell, eds., *Qualitative Researching with Text, Image and Sound*, 227-45. London: Sage.

Pentony, Joseph F. 1998. Effects of negative campaigning on vote, semantic differential, and thought listing. *Journal of Applied Social Psychology* 28(23): 2131-49.

Perloff, Richard M. 2003. *The Dynamics of Persuasion: Communication and Attitudes in the 21st Century*. Mahwah, NJ: Lawrence Erlbaum.

Petrocik, John R. 1996. Issue ownership in presidential elections, with a 1980 case study. *American Journal of Political Science* 40(3): 825-50.

Pettit, Robin T. 2009. Resisting marketing: The case of the British Labour Party under Blair. In J. Lees-Marshment, ed., *Political Marketing: Principles and Applications*, 158-61. New York: Routledge.

Petty, Richard E., and John T. Cacioppo. 1981. *Attitudes and Persuasion: Classic and Contemporary Approaches*. Dubuque, IA: Willam C. Brown.

–. 1986. *Communication and Persuasion: Central and Peripheral Routes to Attitude Change.* New York: Springer-Verlag.

Petty, Richard E., and Duane T. Wegener. 1998. Attitude change: Multiple roles for persuasion variables. In D.T. Gilbert, S.T. Fiske, and G. Lindzey, eds., *The Handbook of Social Psychology,* vol. 1, 323-90. Boston: McGraw-Hill.

Pew Research Center for the People and the Press. 2008. Audience segments in a changing news environment: Key news audiences now blend online and traditional sources. http://people-press.org/.

Pfetsch, Barbara. 2007. Government news management: Institutional approaches and strategies in three Western democracies. In D. Graber, D. McQuail, and P. Norris, eds., *The Politics of News: The News of Politics,* 71-97. Washington, DC: Congressional Quarterly Press.

Phillips, Joan M., Joel E. Urbany, and Thomas J. Reynolds. 2008. Confirmation and the effects of valenced political advertising: A field experiment. *Journal of Consumer Research* 34(6): 794-806.

Phillips, Susan D. 2001. From charity to clarity: Reinventing federal government-voluntary sector relationships. *Philanthropist* 16(4): 240-62.

–. 2006. *The Intersection of Governance and Citizenship in Canada: Not Quite the Third Way.* Ottawa: Institute for Research on Public Policy.

–. 2007. Policy analysis and the voluntary sector: Evolving policy styles. In L. Dobuzinskis, M. Howlett, and D. Laycock, eds., *Policy Analysis in Canada: State of the Art,* 497-522. Toronto: University of Toronto Press.

Pickup, Mark, J. Scott Matthews, Will Jennings, Robert Ford, and Stephen D. Fisher. 2011. Why did the polls overestimate Liberal Democrat support? Sources of polling error in the 2010 British general election. *Journal of Elections, Public Opinion and Parties* 21(2): 179-209.

Pole, Antoinette. 2005. Black bloggers and the blogosphere. Paper presented at the International Conference on Technology, Knowledge and Society, Hyderabad, India, December.

–. 2010. *Blogging the Political.* New York: Routledge.

Politics Watch. 2006. Transcript of meeting between PMO and Press Gallery officials. March 24. http://www.politicswatch.com/.

Pratte, André. 2004. Le blues d'une businesswoman. *La Presse,* January 22, A22.

Pross, Paul, and Kernaghan R. Webb. 2002. Will the walls come tumblin' down? The possible consequences of liberalizing advocacy constraints on charities. Paper presented at the 2002 Public Policy and the Third Sector Conference, Kingston, ON, November.

Proulx, Serge. 2012. L'irruption des médias sociaux: Enjeux éthiques et politiques. In S. Proulx, M. Millette, and L. Heaton, eds., *Médias Sociaux: Enjeux pour la Communication,* 9-31. Montreal: Presses de l'Université du Québec.

Proulx, Serge, and Mary Jane Kwok Choon. 2011. L'usage des réseaux socionumériques: Une intériorisation douce et progressive du contrôle social. *Hermès* 1(59): 105-11.

Public Works and Government Services Canada. 2008. *Trends in Public Opinion Research in the Government of Canada, 2006-7.* http://www.tpsgc-pwgsc.gc.ca/.

–. 2012. *Public Opinion Research in the Government of Canada.* http://www.tpsgc-pwgsc. gc.ca/.

–. 2013. Crown copyright and licensing, integrated services branch. Personal communication, July 30.

Pullman, Emma. 2012. Not planning to vote Conservative? The Tories still want to know all about you. *Vancouver Observer,* March 7. http://www.vancouverobserver. com/.

Pundits' Guide. 2011. The hidden life of parties: A $600 polling budget, son of CIMS?, and PTA finances revealed. http://www.punditsguide.ca/.

Putnam, Robert D. 2000. *Bowling Alone: The Collapse and Revival of American Community.* New York: Simon and Schuster.

Quintelier, Ellen, and Sara Vissers. 2008. The effect of Internet use on political participation: An analysis of survey results for 16-year-olds in Belgium. *Social Science Computer Review* 26(4): 411-27.

Rae, Bob. 1996. *From Protest to Power: Personal Reflections on a Life in Politics.* Toronto: Viking Canada.

Rahat, Gideon, and Tamir Sheafer. 2007. The personalization(s) of politics: Israel, 1949-2003. *Political Communication* 24(1): 65-80.

Rainie, Lee. 2005. Data memo: The state of blogging. http://www.pewinternet.org/.

Ratuva, Steven. 2003. The politics of the media: A cynical synopsis. *Pacific Journalism Review* 9(1): 177-81.

Rawnsley, Gary D. 2005. *Political Communication and Democracy.* Basingstoke, UK: Palgrave Macmillan.

Reber, Bryan H., and Yuhmiin Chang. 2000. Assessing cultivation theory and public health model for crime reporting. *Newspaper Research Journal* 21(4): 99-112.

Reid, Angus. 2013. Angus Reid: Polling isn't broken, but BC shows change is needed. *Globe and Mail,* May 16. http://www.theglobeandmail.com/.

Reid, Elizabeth J. 1999. Non-profit advocacy and political participation. In E.T. Boris and C.E. Steuerle, eds., *Non-Profits and Government: Collaboration and Conflict,* 291-328. Washington, DC: Urban Institute Press.

Reinemann, Carsten, and Jürgen Wilke. 2007. It's the debates, stupid! How the introduction of televised debates changed the portrayal of Chancellor candidates in the German press, 1949-2005. *International Journal of Press/Politics* 12(4): 92-111.

Riddell, Peter. 1998. Members and Millbank: The media and Parliament. *Political Quarterly* 69(B): 8-18.

Ries, Al, and Laura Ries. 2002. *The 22 Immutable Laws of Branding: How to Build a Product or Service into a World-Class Brand.* New York: HarperCollins.

Ritchie, Donald A. 2005. *Reporting from Washington: The History of the Washington Press Corps.* Oxford: Oxford University Press.

Robinson, Gertrude, and Armande Saint-Jean. 1995. The portrayal of women politicians in the media. In Francois-Pierre Gingras, ed., *Gender and Politics in Contemporary Canada,* 176-83. Oxford: Oxford University Press.

Romanow, Walter I., Michel de Repentigny, Stanley B. Cunningham, Walter C. Soderlund, and Kai Hildebrant, eds. 1999. *Television Advertising in Canadian Elections: The Attack Mode, 1993.* Waterloo, ON: Wilfrid Laurier University Press.

Roncarolo, Franca. 2005. Campaigning and governing: An analysis of Berlusconi's rhetorical leadership. *Modern Italy* 10(1): 75-93.

Rose, Jonathan. 2000. *Making Pictures in Our Heads: Government Advertising in Canada.* Westport, CT: Praeger.

–. 2004. Television attack ads: Planting the seeds of doubt. *Policy Options* 25(8): 92-96.

–. 2012. Are negative ads positive? Political advertising and the permanent campaign. In D. Taras and C. Waddell, eds., *How Canadians Communicate IV: Media and Politics,* 149-68. Edmonton: Athabasca University Press.

Rosenberg, Shawn W., Lisa Bohan, Patrick McCafferty, and Kevin Harris. 2001. The image and the vote: The effect of candidate presentation on voter preference. *American Journal of Political Science* 30(1): 108-27.

Roy, Jason, and Nicole Power. 2012. "Cyber-citizens" in the 2011 Canadian federal election: An examination of individual-level social media use. Paper presented at the annual meeting of the Canadian Political Science Association, Edmonton, June.

Roy, Jeffrey. 2006. *E-Government in Canada: Transformation for the Digital Age.* Ottawa: University of Ottawa Press.

Russell, Peter H. 2008. *Two Cheers for Minority Government: The Evolution of Canadian Parliamentary Democracy.* Toronto: Emond Montgomery.

Ryckewaert, Laura. 2012. A year under chief of staff Wright, PMO has less a "bunker mentality." *Hill Times,* May 28. http://www.hilltimes.com/.

Sabatier, Paul A., and Hank C. Jenkins-Smith. 1993. *Policy Change and Learning: An Advocacy Coalition Approach.* Boulder, CO: Westview Press.

Sabato, Larry. 1981. *The Rise of Political Consultants: New Ways of Winning Elections.* New York: Basic Books.

Salamon, Lester M. 1987. *Holding the Center: America's Non-Profit Sector at a Crossroads.* New York: Nathan Cummings Foundation.

Salamon, Lester M., Helmut K. Anheier, Regina List, Stefan Toepler, and Wojciech S. Sokolowski. 1999. *Global Civil Society.* Baltimore: Johns Hopkins University, Center for Civil Society Studies.

Sallot, Jeff. 2006. Canada boosts aid to war-torn Darfur region. *Globe and Mail,* May 24, A5.

Salzman, Jason. 1998. *Making the News: A Guide for Non-Profits and Activists.* Boulder, CO: Westview Press.

Sampert, Shannon. 2012. Verbal smackdown: Charles Adler and Canadian talk radio. In D.T. Taras and C. Waddell, eds., *How Canadians Communicate IV: Media and Politics,* 295-315. Edmonton: Athabasca University Press.

Sampert, Shannon, and Linda Trimble. 2010. Appendix. In S. Sampert and L. Trimble, eds., *Mediating Canadian Politics,* 327-37. Toronto: Pearson Education.

Sanders, Karen. 2009. *Communicating Politics in the Twenty-First Century.* Basingstoke, UK: Palgrave Macmillan.

Sauvageau, Florian. 2012. The uncertain future of the news. In D. Taras and C. Waddell, eds., *How Canadians Communicate IV: Media and Politics,* 29-43. Edmonton: Athabasca University Press.

Savigny, Heather. 2005. Labour, political marketing and the 2005 election: A campaign of two halves. *Journal of Marketing Management* 21(10): 925-41.

Savoie, Donald J. 1999a. *Governing from the Centre: The Concentration of Power in Canadian Politics*. Toronto: University of Toronto Press.

–. 1999b. The rise of court government in Canada. *Canadian Journal of Political Science* 32(4): 635-64.

–. 2010. *Power: Where Is It?* Montreal and Kingston, ON: McGill-Queen's University Press.

Scammell, Margaret. 2007. Political brands and consumer citizens: The rebranding of Tony Blair. *The Annals of the American Academy of Political and Social Science* 611(1): 176-92.

Schenck-Hamlin, William J., David, E. Procter, and Deborah J. Rumsey. 2000. The influence of negative advertising frames on political cynicism and politician accountability. *Human Communication Research* 26(1): 53-75.

Scherer, Jay, and Lisa McDermott. 2011. Playing promotional politics: Mythologizing hockey and manufacturing "ordinary" Canadians. *International Journal of Canadian Studies* 1(43): 107-34.

Scheufele, Dietram A., and David Tewksbury. 2007. Framing, agenda setting, and priming: The evolution of three media effects models. *Journal of Communication* 57(1): 9-20.

Schmid, Hillel, Michal Bar, and Ronit Nirel. 2008. Advocacy activities in non-profit human service organizations: Implications for policy. *Non-Profit and Voluntary Sector Quarterly* 37(4): 581-602.

Schudson, Michael. 1998. *The Good Citizen: A History of American Civic Life*. New York: Free Press.

Scott, D. Travers. 2007. Pundits in muckrakers clothing: Political blogs and the 2004 US presidential election. In M. Tremayne, ed., *Blogging, Citizenship, and the Future of Media*, 39-58. New York: Routledge.

Seawright, David. 2005. "On a low road": The 2005 Conservative campaign. *Journal of Marketing Management* 21(9-10): 943-57.

Sebastian, Michael. 2012. Study: Audiences favor press releases with visuals. *Ragan's PR Daily*, November 29. http://www.prdaily.com/.

Semetko, Holli A. 1996. The media. In L. LeDuc, R.G. Niemi, and P. Norris, eds., *Comparing Democracies: Elections and Voting in Global Perspective*, 254-79. Thousand Oaks, CA: Sage.

Semetko, Holli A., and Margaret Scammell, eds. 2012. *The Sage Handbook of Political Communication*. London: Sage.

Seymour-Ure, Colin. 1962. The Parliamentary Press Gallery in Ottawa. *Parliamentary Affairs* 16(1): 35-41.

Shah, V. Dhavan, Nojin Kwak, and R. Lance Holbert. 2001. Connecting and disconnecting with civic life: Patterns of Internet use and the production of social capital. *Political Communication* 18(2): 141-62.

Shanahan, Elizabeth, Mark McBeth, Paul Hathaway, and Ruth Arnell. 2008. Conduit or contributor? The role of media in policy change theory. *Policy Sciences* 41(2): 115-38.

Shapiro, Robert Y., and John T. Young. 1989. Public opinion and the welfare state: The United States in comparative perspective. *Political Science Quarterly* 104(1): 59-89.

Shirky, Clay. 2011. The political power of social media technology, the public sphere, and political change. *Foreign Affairs* 90(1): 28.

Shoemaker, Pamela J. 1989. *Communication Campaigns about Drugs: Government, Media and the Public*. Hillside, NJ: Lawrence Erlbaum Associates.

–. 1991. *Communication Concepts 3: Gatekeeping*. Newbury Park, CA: Sage.

Shoemaker, Pamela J., and Tim P. Vos. 2009. *Gatekeeping Theory*. New York: Routledge.

Siegel, Arthur. 1993. *Politics and the Media in Canada*. 2nd ed. Toronto: McGraw-Hill Ryerson.

Simpson, Jeffrey. 2001. *The Friendly Dictatorship*. Toronto: McClelland and Stewart.

Skinner, David, James Robert Compton, and Mike Gasher. 2005. *Converging Media, Diverging Politics: A Political Economy of News Media in Canada and the United States*. Oxford: Lexington Books.

Small, Tamara A. 2004. parties@canada: The Internet and the 2004 Cyber-Campaign. In J.H. Pammett and C. Dornan, eds., *The Canadian General Election of 2004*, 203-34. Toronto: Dundurn Press.

–. 2008a. Blogging the Hill: Garth Turner and the Canadian parliamentary blogosphere. *Canadian Political Science Review* 2(3): 103-24.

–. 2008b. Equal access, unequal success: Major and minor Canadian parties on the net. *Party Politics* 14(1): 51-70.

–. 2010a. Canadian politics in 140 characters: Party politics in the Twitterverse. *Canadian Parliamentary Review* 33(3): 39-45.

–. 2010b. Still waiting for an Internet prime minister: Online campaigning by Canadian political parties. In H. McIvor, ed., *Election*, 173-98. Toronto: Emond Montgomery.

–. 2012. E-ttack politics: Negativity, the Internet, and Canadian political parties. In D. Taras and C. Waddell, eds., *How Canadians Communicate IV: Media and Politics*, 169-88. Edmonton: Athabasca University Press.

Smith, Aaron, Kay Lehman Schlozman, Verba Sidney, and Henry Brady. 2009. The Internet and civic engagement. *Pew Internet and American Life Project*. http://pewinternet.org/.

Smith, Gareth. 2005. Politically significant events and their effect on the image of political parties. *Journal of Political Marketing* 4(2/3): 91-114.

Smith, Gareth, and Alan French. 2009. The political brand: A consumer perspective. *Marketing Theory* 9(2): 209-26.

Smith, Joanna. 2010. Is PM's sudden switch to glasses an image makeover? *Toronto Star*, September 5. http://www.thestar.com/.

Smith, Wendell R. 1956. Product differentiation and market segmentation as alternative marketing strategies. *Journal of Marketing* 21(1): 3-8.

Snider, Paul B. 1967. Mr. Gates revisited: A 1966 version of the 1949 case study. *Journalism Quarterly* 27(4): 383.

Solop, Frederic I. 2009. RT @BarackObama we just made history: Twitter and the 2008 Presidential Election. In J.A. Hendricks and R.E. Denton Jr., eds., *Communicator-in-Chief: A Look at How Barack Obama Used New Media Technology to Win the White House*, 37-50. Lanham, MD: Lexington Books.

Soroka, Stuart N. 2002. *Agenda-Setting Dynamics in Canada*. Vancouver: UBC Press.

Soroka, Stuart N., Fred Cutler, Dietlind Stolle, and Patrick Fournier. 2011. Capturing change (and stability) in the 2011 campaign. *Policy Options,* June/July: 70-77.

Soroka, Stuart, and Sarah Robertson. 2010. A literature review of public opinion research on Canadian attitudes towards multiculturalism and immigration, 2006-9. Ottawa: Citizenship and Immigration Canada.

Spears, Tom. 2012. Canadian bureaucracy and a joint study with NASA. *Ottawa Citizen,* April 17. http://www.ottawacitizen.com/.

Spitfire Strategies. n.d. Smart Chart 3.0: An even more effective tool to help nonprofits make smart communication choices. San Francisco: Communications Leadership Institute.

St. Martin, Romeo. 2006. PMO proposes secret cabinet meetings to avoid press scrutiny. *Politics Watch,* March 27. http://www.politicswatch.com/.

Stanley, Woody J., and Christopher Weare. 2004. The effects of Internet use on political participation: Evidence from an agency online discussion forum. *Administration and Society* 36(5): 503-27.

Stanyer, James. 2007. *Modern Political Communications: Mediated Politics in Uncertain Terms.* Cambridge, UK: Polity.

–. 2013. *Intimate Politics.* Cambridge, UK: Polity.

Street, John. 2004. Celebrity politicians: Popular culture and political representation. *British Journal of Politics and International Relations* 6(4): 435-52.

–. 2011. *Mass Media, Politics and Democracy.* 2nd ed. London: Palgrave Macmillan.

Strömbäck, Jesper. 2011. Mediatization and perceptions of the media's political influence. *Journalism Studies* 12(4): 423-39.

Su, Norman M., Yang Wang, and Gloria Mark. 2005. Politics as usual in the blogosphere. Paper presented at the Fourth International Workshop on Social Intelligence Design, Stanford, CA, March. http://www.isr.uci.edu/.

Sundar, S. Shyam, and Clifford Nass. 2001. Conceptualizing sources in online news. *Journal of Communication* 51(1): 52-72.

Sweetser, Kaye D., and Ruthann W. Lariscy. 2008. Candidates make good friends: An analysis of candidates' uses of Facebook. *International Journal of Strategic Communication* 2(3): 175-98.

Sysomos Inc. 2010. Twitter statistics for 2010. An in-depth report at Twitter's growth 2010, compared with 2009. Sysomos Inc. http://www.sysomos.com/.

Takeshita, Toshio. 2006. Current critical problems in agenda-setting research. *International Journal of Public Opinion Research* 18(3): 275-96.

Tapscott, Don. 2009. *Grown up Digital: How the Net Generation Is Changing Your World.* New York: McGraw-Hill.

Taras, David. 1990. *The Newsmakers: The Media's Influence on Canadian Politics.* Scarborough, ON: Nelson.

–. 2001. *Power and Betrayal in the Canadian Media.* Peterborough, ON: Broadview Press.

Taras, David, Fritz Pannekoek, and Maria Bakardjieva, eds. 2007. *How Canadians Communicate II: Media, Globalization and Identity.* Calgary: University of Calgary Press.

Taras, David, and Christopher Waddell, eds. 2012a. *How Canadians Communicate IV: Media and Politics.* Edmonton: Athabasca University Press.

–. 2012b. The 2011 federal election and the transformation of Canadian media and politics. In D. Taras and C. Waddell, eds., *How Canadians Communicate IV: Media and Politics*, 71-107. Edmonton: Athabasca University Press.

Teixeira, Ruy A. 1987. *Why Americans Don't Vote: Turnout Decline in the United States, 1960-1984*. Westport, CT: Greenwood Press.

Tenpas, Kathryn Dunn. 1997. *Presidents as Candidates: Inside the White House for the Presidential Campaign*. New York: Garland Press.

–. 2000. The American presidency: Surviving and thriving amidst the permanent campaign. In T. Mann and N. Ornstein, eds., *The Permanent Campaign and Its Future*, 108-33. Washington, DC: American Enterprise Institute and Brookings Institution.

Tenpas, Kathryn Dunn, and James A. McCann. 2007. Testing the permanence of the permanent campaign: An analysis of presidential polling expenditures, 1977-2002. *Public Opinion Quarterly* 71(3): 349-66.

Theckedath, Dillan, and Terrence J. Thomas. 2012. *Media Ownership and Convergence in Canada*. http://www.parl.gc.ca/.

Thrall, Trevor, Jaime Lollio-Fakhreddine, Jon Berent, Lana Donnelly, Wes Herrin, Zachary Paquette, Rebecca Wenglinski, and Amy Wyatt. 2008. Star power: Celebrity advocacy and the evolution of the public sphere. *International Journal of Press/Politics* 13(4): 362-85.

Tolbert, Caroline J., and Ramona S. McNeal. 2003. Unraveling the effects of the Internet on political participation. *Political Research Quarterly* 56(2): 175-85.

Towner, Terri, and David A. Dulio. 2011. The Web 2.0 election: Does the online medium matter? *Journal of Political Marketing* 10(1): 165-88.

Trimble, Linda. 2005. Who framed Belinda Stronach? National newspaper coverage of the Conservative Party of Canada's 2004 leadership race. Paper presented at the annual meeting of the Canadian Political Science Association, London, ON, June.

–. 2007. Gender, political leadership and media visibility: *Globe and Mail* coverage of Conservative Party of Canada leadership contests. *Canadian Journal of Political Science* 40(4): 976-93.

Trimble, Linda, and Jane Arscott. 2008. *Still Counting: Women in Politics across Canada*. Toronto: University of Toronto Press.

Trimble, Linda, and Joanna Everitt. 2010. Belinda Stronach and the gender politics of celebrity. In S. Sampert and L. Trimble, eds., *Mediating Canadian Politics*, 50-74. Toronto: Pearson Canada.

Trimble, Linda, and Shannon Sampert. 2004. Who's in the game? The framing of the Canadian election 2000 by the *Globe and Mail* and the *National Post*. *Canadian Journal of Political Science* 37(1): 51-71.

Trimble, Linda, Natasia Treiberg, and Sue Girard. 2010. Kim-Speak: L'effet du genre dans la médiatisation de Kim Campbell durant la campagne pour l'élection nationale canadienne de 1993. *Recherches Féministes* 23(1): 29-52.

Trippi, Joe. 2004. *The Revolution Will Not Be Televised: Democracy, the Internet, and the Overthrow of Everything*. New York: Regan Books.

Tulis, Jeffrey K. 1987. *The Rhetorical Presidency*. Princeton, NJ: Princeton University Press.

Turcotte, André. 2012. Under new management: Market intelligence and the Conservative Party's resurrection. In A. Marland, T. Giasson, and J. Lees-Marshment, eds., *Political Marketing in Canada*, 76-90. Vancouver: UBC Press.

Turnbull, Leslie. 2010. Former campaign director, Manitoba NDP. Personal communication, April 23.

–. 2011. Personal communication, September 1.

Twitter. 2011. One hundred million voices. http://blog.twitter.com/.

Van Acker, Elizabeth. 2003. Portrayals of politicians and women's interests: Saviour, sinners, and stars. Paper presented at the annual conference of the Australasia Political Studies Association, Hobart, Tasmania, September.

Van Aelst, Peter, Tamir Sheafer, and James Stanyer. 2011. The personalization of mediated political communication: A review of concepts, operationalizations and key findings. *Journalism* 13(2): 203-20.

Van Onselen, Peter, and Wayne Errington. 2007. The democratic state as a marketing tool: The permanent campaign in Australia. *Commonwealth and Comparative Politics* 45(1): 78-94.

Van Santen, Rosa, and Liesbet van Zoonen. 2009. Popularization and personalization in political communication: A conceptual analysis. Paper presented at the annual conference of the International Communication Association, Chicago, May.

Van Zoonen, Liesbet. 1998. Women and the media: Finally I have my mother back! Politicians and their families in popular culture. *International Journal of Press/Politics* 3(1): 48-64.

–. 2005. *Entertaining the Citizen: When Politics and Popular Culture Converge*. Lanham, MD: Rowman and Littlefield.

–. 2006. The personal, the political and the popular: A woman's guide to celebrity politics. *European Journal of Cultural Studies* 9(3): 287-301.

Van Zoonen, Liesbet, and Christina Holtz-Bacha. 2000. Personalization in Dutch and German politics: The case of the talk-show. *The Public* 7(2): 45-56.

Vastel, Michel. 2003. "L'épine au pied de Paul Martin." *Le Soleil*, 15 February, D7.

Viégas, Fernanda B. 2006. Bloggers' expectations of privacy and accountability: An initial survey. *Journal of Computer-Mediated Communication* 10(3). http://jcmc.indiana.edu/.

Vliegenthart, Rens, Hajo G. Boomgaarden, and Jelle W. Boumans. 2011. Changes in political news coverage: Personalization, conflict and negativity in British and Dutch Newspapers. In K. Brants and K. Voltmer, eds., *Political Communication in Postmodern Democracy: Challenging the Primacy of Politics*, 92-110. New York: Palgrave Macmillan.

Vogt, Paul. 2010. Former clerk of the executive council and cabinet secretary, Government of Manitoba. Personal communication, April 23.

Waddell, Christopher, and Christopher Dornan. 2006. The media and the campaign. In J.H. Pammett and C. Dornan, eds., *The Canadian Federal Election of 2006*, 220-52. Toronto: Dundurn Press.

Wallsten, Kevin. 2005. Blogs and the bloggers who blog them: Is the political blogosphere an echo chamber? Paper presented at the annual meeting of the American Political Science Association, Washington, DC, September.

Walters, Richard D., and Weija Wang. 2011. Painting a picture of America's non-profit foundations: A content analysis of public relations photographs distributed through online wire services. *International Journal of Non-Profit and Voluntary Sector Marketing* 16(2): 138-49.

Wanta, Wayne. 1997. *The Public and the National Agenda: How People Learn about Important Issues.* Mahwah, NJ: Lawrence Erlbaum Associates.

Ward, Ian. 2002. The Tampa, wedge politics, and a lesson for political journalism. *Australian Journalism Review* 24(1): 21-39.

Ward, Stephen, and Rachel Gibson. 2009. European political organizations and the Internet: Mobilization, participation, and change. In A. Chadwick and P.N. Howard, eds., *The Routledge Handbook of Internet Politics*, 25-39. Abingdon, UK: Routledge.

Ward, Stephen, Rachel Gibson, and Paul Nixon. 2003. Parties and the Internet: An overview. In R. Gibson, P. Nixon, and S. Ward, eds., *Political Parties and the Internet Net Gain?* 11-39. London: Routledge.

Waters, Richard D., Emily Burnett, Anna Lamm, and Jessica Lucas. 2009. Engaging stakeholders through social networking: How non-profit organizations are using Facebook. *Public Relations Review* 36(2): 102-6.

Waters, Richard D., Natalie T.J. Tindall, and Timothy S. Morton. 2010. Media catching and the journalistic-public relations practitioner relationship: How social media are changing the practices of media relations. *Journal of Public Relations Research* 22(3): 241-64.

Watt, Laurie. 2013. PM's office sends financial details of Trudeau speech to newspaper. *Barrie Advance*, June 18. http://www.thestar.com/.

Weaver, David H. 1977. Political issues and voter need for orientation. In D. Shaw and M.E. McCombs, eds., *The Emergence of American Political Issues: The Agenda-Setting Function of the Press*, 107-19. St. Paul, MN: West Publishing.

Webb, Kernaghan. 2000. *Cinderella's Slippers? The Role of Charitable Tax Status in Financing Canadian Interest Groups.* Vancouver: SFU Centre for the Study of Government and Business.

Weber, Lori M., Alysha Loumakis, and James Bergman. 2003. Who participates and why? An analysis of citizens on the Internet and the mass public. *Social Science Computer Review* 21(1): 26-42.

Wei, Ran, and Lo Ven-Hwei. 2007. The third-person effects of political attack ads in the 2004 US presidential election. *Media Psychology* 9(2): 367-88.

Wesley, Jared J. 2006. The collective centre: Social democracy and red Tory politics in Manitoba. Paper presented at the annual meeting of the Canadian Political Science Association, Toronto, May.

–. 2011a. *Code Politics: Campaigns and Cultures on the Canadian Prairies.* Vancouver: UBC Press.

–. 2011b. Staking the progressive centre: An ideational analysis of Manitoba party politics. *Journal of Canadian Studies* 45(1): 143-77.

Wesley, Jared J., and Wayne Simpson. 2011. Promise meets reality: Balanced budget legislation in western Canada, 1990-2010. Paper presented at the annual meeting of the Canadian Political Science Association, Waterloo, ON, May 16.

Wesley, Jared J., and David Stewart. 2006. Campaign finance reform in Manitoba: Advantage Doer? Paper presented at the Conference on Party and Election Finance, Calgary, May.

West, Darrell M. 2009. *Air Wars: Television Advertising in Election Campaigns, 1952-2008.* 5th ed. Washington, DC: Congressional Quarterly Press.

Wherry, Aaron. 2011. Why Harper is never in cellphone range. *Maclean's,* July 11. http://www2.macleans.ca/.

-. 2012. Every day is Election Day in Canada. *Maclean's,* January 9. http://www2.macleans.ca/.

White, David Manning. 1950. The "gate keeper": A case study in the selection of news. *Journalism Quarterly* 17(4): 383-90.

White, Jon, and Leslie de Chernatony. 2002. New Labour: A study of the creation, development, and demise of a political brand. *Journal of Political Marketing* 1(2): 45-52.

Whitehorn, Alan. 2006. The NDP and the enigma of strategic voting. In J. Pammett and C. Dornan, eds., *The Canadian General Election of 2006,* 93-121. Toronto: Dundurn Press.

Winter, James, Chaim Eyal, and Ann Rogers. 1982. Issue-specific agenda setting: The whole as less than the sum of the parts. *Canadian Journal of Communication* 8(2): 1-10.

Wiseman, Nelson. 1985. *Social Democracy in Manitoba: A History of the CCF/NDP.* Winnipeg: University of Manitoba Press.

Wood, Lisa. 2000. Brands and brand equity: Definition and management. *Management Decision* 38(9): 662-69.

Wright, Gerald C. Jr. 1977. Racism and welfare policy in America. *Social Science Quarterly* 57(4): 718-30.

Wright, Peter L. 1973. The cognitive processes mediating acceptance of advertising. *Journal of Marketing Research* 10(1): 53-62.

Wring, Dominic. 2005. *The Politics of Marketing the Labour Party.* New York: Palgrave Macmillan.

Xenos, Michael, and Patricia Moy. 2007. Direct and differential effect of the Internet on political and civic engagement. *Journal of Communication* 57(4): 704-18.

Yagade, Aileen, and David M. Dozier. 1990. The media agenda-setting effect of concrete versus abstract issues. *Journalism Quarterly* 67(1): 3-10.

Yang, H.S. 2008. The effects of the opinion and quality of user postings on Internet news readers' attitudes toward the news issue. *Korean Journal of Journalism and Communication Studies* 52(2): 254-80.

Yeon, Hye Min, Youjin Choi, and Spiro Kiousis. 2005. Interactive communication features of nonprofit organizations' webpages for the practice of excellence in public relations. *Journal of Website Promotion* 1(4): 61-83.

Young, Lisa, and William Cross. 2002. The rise of plebiscitary democracy in Canadian political parties. *Party Politics* 8(6): 673-99.

Zaller, John. 1992. *The Nature and Origins of Mass Opinion.* Cambridge: Cambridge University Press.

Zamora Medina, Rocio. 2012. Campaigning on Twitter: The use of the "personal style" to activate the political participation in the Spanish elections 2011. Paper presented at the twenty-second World Congress of Political Science, Madrid, July 8-12.

Zerbisias, Antonia. 2006. Taking the Hill. Blog. http://thestar.blogs.com/.

Zucker, Harold G. 1978. The variable nature of news media influence. In B. Ruben, ed., *Communication Yearbook 2*, 225-40. New Brunswick, NJ: Transaction Books.

Contributors

PÉNÉLOPE DAIGNAULT (Université Laval) has published about social advertising and the measure of its effectiveness, fear appeals, the role of empathy in persuasion processes, and, most recently, Canadian political advertising. Her research interests mainly revolve around the impact of social and political advertising.

SUSAN DELACOURT *(Toronto Star)* has been reporting on Canadian politics since the late 1980s and has written four books. Her latest, *Shopping for Votes*, looks at how consumer culture has mixed with Canadian political culture.

ALEX DROUIN (Université du Québec à Trois-Rivières) completed a master's degree in communication at the Université du Québec à Trois-Rivières about the gendered representations of political actors in editorial cartoons. His research interests include the image of politicians and gender framing.

ANNA ESSELMENT (University of Waterloo) has published about parties, elections, and partisanship in *Canadian Journal of Political Science, Canadian Public Administration,* and *Publius: The Journal of Federalism.* Her research interests include political parties, campaigns and elections, Canadian institutions, and the role of partisanship in intergovernmental relations in Canada.

THIERRY GIASSON (Université Laval) has published about media hypes, political marketing, televised leaders debates, and political bloggers. He was co-editor, with Alex Marland and Jennifer Lees-Marshment, of *Political Marketing in Canada* (UBC Press, 2012). His research interests include online political campaigns, political journalism, biopolitics, Quebec politics, and democratic malaise in Canada.

ELISABETH GIDENGIL (McGill University) has published widely on issues relating to voting behaviour, public opinion, and political communication. She was a member of the Canadian Election Study team from 1992 to 2008 and was principal investigator for the 2008 study.

GEORGINA GROSENICK (Carleton University) has researched and analyzed the strategic communication and advocacy practices of non-profit organizations. Gina worked for twenty years in the non-profit sector, primarily with trade and industry organizations, as a lobbyist, communicator, executive, and project manager. She teaches in the areas of strategic/critical communications and media analysis.

HAROLD JANSEN (University of Lethbridge) has published on Canadian political party finance reform, Alberta politics, web-based campaigning, blogging, Internet discussion boards, and electoral reform in Canada's provinces. His current research focus is the impact of the use of digital communications on democratic citizenship in Canada.

ROYCE KOOP (University of Manitoba) authored *Grassroots Liberals: Organizing for Local and National Politics in Canada* (UBC Press, 2011), as well as articles in *Canadian Journal of Political Science, Representation,* and *Journal of Elections, Public Opinion, & Parties.* His research focuses on political parties, federalism, and Canadian politics.

MIREILLE LALANCETTE (Université du Québec à Trois-Rivières) has published about gender, media, and representation. Her research interests include political communication, media, and representations, with a particular emphasis on framing. She is currently working on the transformations of political actors' representations in the context of spectacularization and personalization in the media.

ANDREA LAWLOR (University of California, Berkeley) is a postdoctoral fellow specializing in Canadian and comparative public policy. Her research interests include immigration and social policy, public opinion, media analysis, and Canadian political institutions. Andrea's research can be found in *Comparative and Commonwealth Politics*, *Canadian Public Administration*, and *Electronic Media and Politics*.

CATHERINE LEMARIER-SAULNIER (Université Laval) is a doctoral student in public communication at Université Laval. She has published about political coverage of female politicians in *Canadian Journal of Communication*. Her research interests include gender, media, and politics.

ADAM MAHON (McGill University) is with the Institute for Health and Social Policy, specializing in cross-national comparisons of poverty reduction policies. His research interests include comparative social security systems, Canadian public opinion, and media content analysis.

ALEX MARLAND (Memorial University) was the lead editor of *Political Marketing in Canada* (UBC Press, 2012) and *First among Unequals: The Premier, Politics, and Public Policy in Newfoundland & Labrador* (MQUP, 2014). His research interests include political communication, electioneering, and public policy in Canada.

J. SCOTT MATTHEWS (Memorial University) is a specialist in the study of elections, voting, and public opinion in Canada and the United States. His past work includes papers published in *British Journal of Political Science*, *European Journal of Political Research*, *Electoral Studies*, and *Canadian Journal of Political Science*.

DENVER MCNENEY (McGill University) is a doctoral student at the Centre for the Study of Democratic Citizenship. His research links work in cognitive science, political communication, and social psychology to connect the range of individual and environmental processes that help form citizens' attitudes.

MIKE MOYES (University of Manitoba) is completing a master's of arts thesis titled "Doer and Dexter: A Model for the NDP." His research interests include political marketing, Canadian politics, and public administration.

DANIEL J. PARÉ (University of Ottawa) has published about political marketing, information, and communication policy, social justice, and Internet governance. His research interests focus on social, economic, political, and technical issues arising from innovations in information and communication technologies in transitioning and industrialized countries.

TAMARA A. SMALL (University of Guelph) has published work on online election campaigning in *Party Politics* and in *Canadian Journal of Political Science,* and about Internet regulation in *Election Law Journal.* Her research interests include digital politics in Canada, political communications, and party politics.

STUART SOROKA (McGill University) is a William Dawson Scholar, a member of the Centre for the Study of Democratic Citizenship, and co-investigator of the Canadian Election Study. His research focuses on interactions among public opinion, public policy, and mass media.

JARED J. WESLEY (University of Alberta) researches Canadian provincial party politics and elections. He is the author of *Code Politics: Campaigns and Cultures on the Canadian Prairies* (UBC Press, 2011), and principal investigator of the SSHRC-funded Comparative Provincial Election Study.

Index

Printed and bound in Canada by Friesens
Set in Scala and Minion by Artegraphica Design Co. Ltd.
Copy editor: Judy Phillips
Proofreader: Lana Okerlund